Physical Hazards of the Workplace

Second Edition

T0262787

Occupational Safety and Health Guide Series

Series Editor

Thomas D. Schneid
Eastern Kentucky University
Richmond, Kentucky

Published Titles

Forthcoming Titles

Safety and Human Resource Law for the Safety Professional,
Thomas D. Schneid

Human Resources & Change Management for Safety Professionals,
Thomas D. Schneid and Shelby L. Schneid

Physical Hazards of the Workplace

Second Edition

Barry Spurlock

CRC Press
Taylor & Francis Group
Boca Raton London New York

CRC Press is an imprint of the
Taylor & Francis Group, an **informa** business

CRC Press
Taylor & Francis Group
6000 Broken Sound Parkway NW, Suite 300
Boca Raton, FL 33487-2742

First issued in paperback 2023

© 2018 by Taylor & Francis Group, LLC
CRC Press is an imprint of Taylor & Francis Group, an Informa business

No claim to original U.S. Government works

ISBN 13: 978-1-03-256968-0 (pbk)
ISBN 13: 978-1-4665-5703-1 (hbk)

DOI: 10.1201/9781315374413

Library of Congress Cataloging-in-Publication Data
Names: Spurlock, Barry, author.
Title: Physical hazards of the workplace / Barry Spurlock.
Description: Second edition. | Boca Raton : Taylor & Francis, CRC Press, 2017. | Series: Occupational safety & health guide series | Includes bibliographical references and index.
Identifiers: LCCN 2017016057| ISBN 9781466557031 (hardback) | ISBN 9781315374413 (ebook)
Subjects: LCSH: Industrial safety--United States.
Classification: LCC T55 .S74 2017 | DDC 658.4/08--dc23
LC record available at https://lccn.loc.gov/2017016057

Publisher's Note
The publisher has gone to great lengths to ensure the quality of this reprint but points out that some imperfections in the original copies may be apparent.

Visit the Taylor & Francis Web site at
http://www.taylorandfrancis.com

and the CRC Press Web site at
http://www.crcpress.com

Contents

Preface

It has been more than one-and-a-half decades since the first edition of this book was published in 2001. When it was released, the cell phone's utility was simply that of making a phone call. Texting and taking pictures with one's mobile phone would not catch on for another two years. At that time, ergonomics and behavior-based safety were hot new topics on the precipice of becoming the central focus of most organizations' safety efforts. In fact, *ergonomics* was deemed a buzzword in the first edition. Also in 2001, most leadership positions within the safety profession were held by those from the baby boomer generation (depending upon the source, this refers to individuals born between the post–World War II 1940s and the mid-1960s).

A lot has changed since the first edition. Now, designing work stations and safety management systems with the human interface in mind is no longer a new idea, but rather a fundamental expectation of competent safety professionals. A canvas of safety leaders within organizations today would reveal a rather complex mix of baby boomers, generation Xers (born between the mid-1960s and mid-1970s), and even some millennials. Along with this blending of generations has come a new era in safety with new thought leaders who ask "why" and carefully examine the value of safety management practices commonly thought of as unchangeable truths. As a result, best practices in safety management have emerged that have revolutionized how occupational safety and health is managed. These best practices have also solidified safety's spot in the C-suite of many companies. In addition to the notable advances in safety management, technology and regulations have changed at a pace matched only by the time period immediately following the passage of the Occupational Safety and Health Act of 1970.

Even though there have been significant developments in occupational safety and health since the first edition of this book, the essence of the profession has not changed. Invariably, the OSH profession still involves four tasks: Anticipating, identifying, analyzing, and controlling hazards in the workplace.* While safety professionals strive to continually hone their skills in performing these tasks, the tasks themselves are timeless and perpetual. The measure of an organization's safety success is marked more by how well it performs these fundamental tasks than by its reduction in injury rates. Injury rates can improve through mere luck, but risk is only reduced through purposeful, quality efforts in identifying and controlling hazards. Successful, continuous efforts to reduce risk will result in long-term, sustainable reductions in occupational injuries and illnesses.

In order to successfully identify and control hazards, organizations must (1) have a sound, systematic process and (2) possess or obtain the requisite, technical expertise about particular hazards. In this second edition, both of these requirements are addressed. New chapters have been included on the systematic process for

---

* Throughout this book, these four tasks simply may be referred to as "identifying and controlling hazards." In no way is this meant to diminish the importance of the anticipation and assessment of hazards.

identifying, analyzing, and controlling hazards in the workplace. This additional information not only helps organizations and professionals more effectively manage their efforts in identifying and controlling hazards, it helps them address one of the most critical elements of a modern occupational health and safety management system as required by consensus standards such as ANSI Z10 and ISO 45001.

As with the first edition, the book still contains chapters that provide valuable information on particular workplace hazards and their controls, but these chapters have been updated and expanded to address regulatory changes and widely accepted best practices. To maximize success and utility with this book, the reader is encouraged to apply the more technical guidance provided in these chapters within the framework of a sound hazard identification and control system as set forth in the beginning chapters.

Whether used as a college textbook or safety practitioner reference, readers should know this book is designed to provide a broad base of information on some of the most prevalent workplace hazards. In many cases, an entire book would be necessary to exhaustively address some of the hazards in all of the contexts/industries in which they exist. While the chapters on systemically identifying and controlling hazards is more exhaustive in nature, each organization should create or adapt their systems with their organization's culture and other business systems in mind. In sum, readers should use this book to develop frameworks for identifying and controlling workplace hazards, and then remain diligent in researching best practices as they pertain to the circumstances and industries in which they work.

The review questions at the end of the chapters are instructive for both the occupational safety student and safety practitioner. Students and faculty should look to the review questions to promote reflection and feedback from the chapters, and gauge comprehension of the material. Safety practitioners can use the questions to identify gaps when conducting training needs assessments, and even facilitate discussions in safety committee and management meetings. Most importantly, the author hopes the questions prompt all readers to pause, think, and memorialize how they can improve their skills and systems for hazard identification and control.

It is the author's desire that this book serve as a valuable tool for readers wishing to make a notable, sustainable difference in identifying and controlling hazards within their organizations. Ultimately, the hope is that this book becomes one of many that makes a remarkable difference in the safety and health of workers. If any single person's mother, father, son, daughter, sibling, or grandchild avoids injury or illness from the work they perform as result of this book, then the efforts were worth it. Godspeed in all your safety initiatives!

Acknowledgments

I am extremely grateful for the support and patience of the staff at Taylor & Francis Publishing Group during the preparation of this manuscript—particularly Cindy Carelli and Renee Nakash. Both of these ladies have proven to be some of the most understanding, patient, and helpful people in the world.

It is sometimes surreal when I think about how I now have the opportunity to teach alongside many of my former professors at Eastern Kentucky University. It is even more amazing for me to consider the fact that I have had the opportunity to be the editor for the second edition of a book that two of them wrote. To my former professors and authors of the first edition, Dr. Tom Schneid and Dr. Larry Collins, I am truly grateful for your mentorship and friendship through the years and in laying the groundwork for this second edition.

I am extremely grateful to Dr. Earl Blair, who was my first professor in this field and the person most responsible for my pursuing a career in occupational safety and health. I thank you for giving me the opportunity to begin my teaching career as an adjunct professor at Indiana University and for your unwavering friendship and mentorship through the years. Many doors in this profession would remain unopened were it not for you.

To my wonderful parents, Bobby and Marilyn, thank you for the sacrifices you made throughout my life to ensure I was the best I could be, and for instilling in me the work ethic and discipline it takes to author books while simultaneously juggling the demands of a full-time profession and large, busy family. None of my success is realized without your influence and support throughout the years. Your love for your son is priceless and unmatched. I can only hope I love my children as much as you have loved me.

To my wife Candace and my children—Emma, Emory, Amber, Alaina, and Evan—I thank you with all that is in me for your support and for making my life what it is today. I know that your sacrifice of time with me in order for me to work on projects such as this are significant. I sincerely hope I make you proud, and that you know how much you are loved. It's my hope that my legacy and example will produce even greater things in each of you. I love you with all my heart.

Most importantly, I am thankful to God Almighty for his grace and the providence of opportunities with which he has provided me, and for blessing me with the abilities, resources, and support to complete a work such as this. It is my hope and prayer that I never take for granted how blessed I am.

About the Author

Barry S. Spurlock is an assistant professor at Eastern Kentucky University in the College of Justice and Safety, and the managing member attorney with Spurlock Law, PLLC. In addition to being a practicing attorney, he is also a board-certified safety professional. At EKU, Spurlock teaches graduate and undergraduate courses on topics such as workers' compensation and labor law, workers' compensation management, occupational safety management, safety performance measurement, and workplace hazard identification and control. Prior to being a full-time faculty member at EKU, Spurlock practiced employment law full-time and devoted a significant amount of his practice to representing and advising employers on OSHA matters. Before becoming an attorney, he worked for more than 16 years as a safety and environmental professional in the steel, food, and insurance industries. He also served as adjunct faculty for Indiana University for more than a decade, and while there he developed one of the first courses on safety performance measurement to be offered at any university. Spurlock holds a Bachelor of Science in industrial safety and risk management, and a Master of Science in loss prevention and safety administration, both from Eastern Kentucky University. He also holds a juris doctor from Northern Kentucky University's Salmon P. Chase College of Law. He is an active member of ASSE and currently serves the Louisville Chapter as a past president and as chair of the Government Affairs Committee. He is also a member of ASSE's National Government Affairs Committee. Spurlock has authored articles for the American Bar Association Tort and Insurance Practice Specialty's Worker's Compensation and Employer's Liability periodical, and is coauthor of the book *Chomp Comp: The Small Business Guide to Lower Worker's Compensation Premiums*. He is a frequent speaker at the most prestigious national safety conferences. Spurlock and his wife Candace have been married for nearly 20 years and have five children: Emma, Emory, Amber, Alaina, and Evan. Spurlock is active in coaching his kids' baseball and basketball teams, and enjoys time in the outdoors with all of them. When not working or spending time with the kids, Barry enjoys playing bluegrass music, fishing, and working on the family's mini-farm.

About the Authors
of the First Edition

Tom Schneid currently serves as chair and tenured professor in EKU's Department of Safety and Security. He has extensive work experience in the safety, human resource, and legal fields, and has represented numerous corporations and individuals in OSHA and labor/employment-related litigations throughout the United States. Dr. Schneid earned a BS in education, MS and CAS in safety, as well as his juris doctor in law from West Virginia University. He later earned his LLM in labor and employment law from the University of San Diego. He is a member of the bar for the U.S. Supreme Court, 6th Circuit Court of Appeals and a number of federal districts as well as the Kentucky and West Virginia Bar. Dr. Schneid has authored or coauthored numerous texts on occupational safety and the legal issues surrounding it.

Larry Collins retired as associate dean in EKU's College of Justice and Safety in 2014. Before his retirement and service as associate dean, Dr. Collins served as a faculty member in EKU's School of Safety Security and Emergency Management, program coordinator of the Fire and Safety Engineering Technology Program, and chair of the department of Loss Prevention and Safety Administration. Dr. Collins holds an AS degree in fire science from Allegheny Community College in Pennsylvania, a BS in industrial arts education from California University of Pennsylvania, and an MS in technology education, also from California University of Pennsylvania.

1 Systematically Managing Hazards in the Workplace
A Framework for Hazard Identification and Control

At the core of all organizations' safety initiatives is the control of hazards. Fundamentally, the occupational safety and health professional's job is to anticipate, identify, analyze, and control hazards.* To maximize the chances of success, these tasks should occur purposefully and within the framework of a sound hazard identification and control system. Just as important as the technical expertise in identifying hazards and designing controls is the ability to accurately assess the risk associated with a hazard and evaluate the effectiveness of controls once they are implemented. This chapter focuses on the later ability to manage the hazard identification and control processes, while the remainder of the book deals with more technical information pertaining to specific hazards.

DEFINING HAZARD AND RISK

After just a short time reviewing the body of literature on occupational safety and health, one will find the terms *hazard* and *risk* are frequently and erroneously used interchangeably. In many instances, the context in which the particular term is used will permit the reader to understand what the writer actually meant, and the lack of precision in language is of no consequence. However, it is important for safety professionals to make distinctions in terminology as they develop a systematic approach to managing hazards in the workplace, and certainly when teaching the next generation of safety professionals. The terms *hazard* and *risk* are indeed distinct and a clear definition of both is warranted at the outset of this book.

HAZARD

A *hazard* is best defined as an object, condition, substance, process, action, or behavior that is a source of potential harm.

---

* Throughout this book, the tasks of anticipating, identifying, analyzing, and controlling hazards will be collectively referred to as hazard identification and control.

In the occupational safety and health (OSH) context, a hazard is essentially anything that has the potential to cause a worker to sustain an injury or illness without respect to the severity or character of the injury/illness, or the likelihood the injury/illness will occur. Commonly, the term *hazard* follows another word or is coupled with a phrase, for instance, laceration hazard, fall hazard, repetitive trauma hazard. This coupling of terms helps specify either a source of harm or the anticipated injury associated with the source. For instance, the term *fall hazard* may be used when evaluating a work platform in a manufacturing facility that lacks a proper handrail. The unguarded platform has the potential to lead to a fall. Should the worker fall, she or he may experience a variety of injuries, or if lucky, no injury at all. One may say the fall (or the sudden stop) is the source of injury; however, it is the unguarded platform that is the source of the fall, and thus "fall hazard" is still appropriate. Conversely, the general public, mass media, and sadly many within the safety profession too often use the term *safety hazard* when they simply mean a hazard or a particular source of a particular harm. In some respects, safety hazard, as it is commonly used, is an oxymoron because the term *safety* is defined as the freedom or absence of harm, and anything that would lead to the freedom of harm is actually desirable.

This book is largely focused on *physical hazards*. To describe a physical hazard, it may be instructive to first explain what a physical hazard is not. Physical hazards are in a large, distinct category of workplace hazards primarily involving sources of harm that exclude biological agents or chemicals. Biological and chemical exposures in the workplace present unique challenges, and quite often require the specialized work of an industrial hygienist to properly identify and assess their presence. Examples of biological agents may include mold, bacteria, viruses, and so on. Chemical hazards affect the body in a variety of routes of entry: inhalation, absorption, ingestion, and injection. Each route of entry requires a unique scientific or medical monitoring method to evaluate worker exposure to the chemical.

Many authorities would describe physical hazards as those hazards that can cause injury without necessarily coming in contact with the body. Important to remember is the word *necessarily*, as there are many physical hazards, such as handling or coming in contact with sharp surfaces, that can cause the body harm (i.e., a laceration). While many physical hazards produce more acute injuries such as lacerations, amputations, and strains/sprains, there are physical hazards that produce more latent injuries, such as carpal tunnel syndrome, hearing loss, and so on. Physical hazards more broadly include sources of harm such as electricity, heat, vibration, noise, contact with equipment and surfaces, slips/trips/falls, and radiation, to name a few. Some common injuries associated with physical hazards include lacerations, punctures, strains/sprains, electrical shock, heat exhaustion/stroke, and repetitive trauma injuries. While some may exclude topics such as bloodborne pathogens and workplace violence from a book on physical hazards, they have been included in this book because the methods of hazard assessment and control parallel that of many physical hazards, and they are issues the vast majority of general safety practitioners must address.

RISK

Risk is an expression that quantifies (1) the probability a negative outcome will occur as a result of a hazard and (2) the potential severity or impact of the negative outcome should it occur.

In order to effectively manage hazard control efforts, one must be able to assess the potential impact a hazard presents to an organization, both before and after control efforts. In other words, once a hazard is identified, a risk assessment must be performed. This assessment involves some type of calculation (either quantitatively or qualitatively) of the degree of risk a particular hazard presents to the organization. Efforts that simply inventory workplace hazards (hazard identification) should not be confused with risk assessment. Risk assessment is a more strategic process that at least considers two variables: likelihood (or frequency) of occurrence and severity. More detailed information concerning risk assessment is provided later in this chapter in the section on assessing hazards.

Accurately defining the terms *hazard* and *risk* is not just an exercise in semantics. Being able to accurately distinguish the two is critical for those charged with developing a hazard identification and control system. One must understand the role of each, and it is an important point of distinction.

ANTICIPATING AND IDENTIFYING HAZARDS

Anticipating and identifying hazards is a prerequisite to effectively analyzing hazards and prescribing the means by which they can be controlled. The hope and goal of the safety professional is to be proactive and identify hazards before an injury occurs. But achieving zero risk is unreasonably utopic for most organizations, and unfortunately an employee's injury may serve as the means of hazard identification. Provided the organization has a robust incident investigation process, identifying hazards in the wake of an injury is generally not a daunting task—in many instances the hazards are now blatantly obvious. (This is not to say that conducting the investigation is easy.) While identifying how multiple hazards (i.e., behavioral and environmental) worked in concert to produce a harm can sometimes be a more challenging task, but the astute investigator can still create a rather comprehensive examination of hazards as a by-product of her or his investigation. Organizations should:

1. Strive to minimize the reactive identification of hazards.
2. Remain extremely diligent and even more aggressive in addressing hazards in the wake of an injury.
3. Aggressively focus on identifying hazards before an incident occurs.

Anticipating and identifying hazards before an incident occurs is often a more daunting task, and it demands foresight. Those within the organization must evaluate the workplace and work processes with the human interface in mind, and continuously perform a "what if" analysis. Further complicating the challenge of proactively identifying hazards is the absence of a negative incident involving the process or condition, and this may lead many workers to dismiss an identified hazard as being far-fetched or unrealistic; thus, they ignore it. On countless occasions the author has heard members of management in several companies state, "No one would ever do that" when discussing a hazard that has been identified. Too many times it has taken an injury for that member of management to concede the hazard was real and presented a realistic, actual risk.

Aside from an incident investigation and injury logs, there are several means by which the safety professional can anticipate and identify hazards before injuries occur. The following represents a non-exhaustive list of these means:

- Informal, verbal employee reports (probably the most common)
- Formal hazard reporting cards/reports
- Injury and illness trend data
- Safety suggestion systems
- Work environment inspections
- Equipment inspections
- Preventive maintenance records
- Prejob/pretask hazard analyses
- Job safety analyses
- Safety meeting discussions/minutes
- Safety climate/perception surveys
- Behavioral observations
- Equipment design drawings and specifications (best utilized in
- Insurance loss control reports (property and workers' compensation)

The above list is certainly not exhaustive, and many means of hazard identification are particular and unique to any given organization. Probably one of the most common and effective means to identify hazards is by having meaningful conversations with the employees who directly interact with the working environment, processes, equipment, procedures, and other employees. A key, however, to hazard identification through direct employee engagement and conversation is an effective means to memorialize the hazard identification and track progress. Employers must be aware, however, that emails, memorandum, and so on, wherein an agent of the employer (supervisor, manager, etc.) documents the hazard are discoverable in litigation, and following up to correct the hazard is imperative. Documenting hazards demonstrates knowledge of the hazard, and in some cases the email, memo, and so on can satisfy the evidentiary burden of proving a willful OSHA violation if the hazard is not properly addressed by the employer. Memorializing reports of hazards does create liability concerns with OSHA and ordinary tort law, but employers should know that diligent efforts to manage the follow-up to hazard reports can reduce liability risks and help reap the safety management benefits of tracking identified hazards.

In many instances, various employees in an organization must engage in the task of purposeful and intentional hazard identification; it cannot simply be the function of the safety coordinator/manager/professional. Accordingly, those engaging in the task of hazard identification must be properly equipped with the requisite knowledge and skills to effectively identify hazards. Simply because an employee has worked at the organization for a long period of time, is capable of supervising other employees, or is very interested in serving on the safety committee does not impart the ability to spot hazards.

This author firmly believes that training is too often thought of as a panacea for all safety problems. Unfortunately, training is not the solution to everything, and in many cases coaching is more desirable than training. However, for those with little to no experience or education in occupational safety, training in hazard identification and assessment is usually necessary. Optimally, an organization can design and conduct customized hazard identification training that includes provisions for working within the organization's hazard identification and control system. If such customized training is not an option, organizations should at least consider having those inexperienced and uneducated personnel attend general training through some recognized outlet (i.e., state/regional safety conference, a local 30-hour general industry or construction safety course through an authorized OSHA outreach training institute/center). Regardless of how hazard identification training is delivered, post-training evaluation and coaching is critical to maximizing the performance of those charged with hazard identification.

ANALYZING HAZARDS (ASSESSING THE RISK)

Once a hazard has been identified, a purposeful effort to assess and quantify the risk associated with the hazard should occur. As stated in the foregoing definitions of this chapter, this assessment involves an attempt to closely examine the hazard and assign a risk quantification based on the likelihood an injury will occur as a result of exposure to the hazard, and what the potential severity and impact of such an injury would be. Once the risk is quantified it should be examined through the established risk tolerance lens of the organization. This final examination helps the organization make decisions ranging from completely abandoning a process (when risk is too high even in the face of possible control methods) to do nothing about the hazard (when no risk or only negligible risk is present). In the vast majority of cases, organizations do not abandon the process and they must do something to reduce the risk associated with the hazard.

Assessing and quantifying risk has traditionally considered two variables: Likelihood and severity. In other words, (Likelihood) × (Severity) = Risk. Likelihood simply captures the chance the negative event will occur, and severity simply accounts for how severe the injury would be—usually expressed as either cost *or* nature of injury. In recent years, however, safety professionals have realized the limitations of such a simplistic view of risk assessment. In response, several variations of the traditional risk calculation formula have popped up. With respect to the likelihood variable, many have identified the dichotomy that exists between how likely an injury is to occur and the frequency at which the task is performed wherein the hazard exists. In its occupational safety and health risk formula, Blair & Spurlock, LLC, even adds a third variable to the likelihood formula that accounts for the percentage of the workforce that actually engages in the task. Similarly, some modern risk assessment formulas have separate variables for injury/harm costs and the character/nature of loss. Blair & Spurlock have also added a third variable in its OSH risk formula's calculation of severity by considering OSHA penalty risks. While risk formulas could quickly and conceivably become too complex and unworkable, evaluating as many facets of risk as possible will help in providing a more precise assessment of the risk.

When designing a risk assessment system and formula, numerical risk score values correspond with some narrative description of the particular aspect of risk. Generally, the higher numeric risk values represent the greater the contribution to the overall risk score. For example, when describing likelihood, a risk score system may equate the highest numerical value of five (5) to mean "occurrence is imminent and certain," while the numerical value of one (1) means "the event is possible but highly unlikely." Similar to severity, a highest numerical value of five (5) may mean "death or disabling injury," while a one (1) may mean "minor first aid injury is the worst-case scenario." As the reader might expect, each variable in a risk formula must have a corresponding narrative description that guides the user in selecting a numerical risk value. Consistency, precision, and just the right amount of detail is a must when establishing a risk assessment system.

An exhaustive discussion of risk assessment is beyond the scope of this book as it is an aspect of safety that pervades nearly all of its subdisciplines, for instance, system safety analysis. It is so exhaustive that organizations such as the American Society of Safety Engineers have established entire institutes and professional development conference tracks on the topic. Safety professionals should dive deep and research the best custom options for establishing a risk assessment system within her or his organization.

CONTROLLING HAZARDS

A book on the subject of physical hazards would not be complete without a fundamental discussion of the framework in which safety professionals have been controlling hazards for decades—the Hierarchy of Controls. The hierarchy is a way to express the categorical methods of controlling hazards and the effectiveness or desirability of each. Essentially, employers control hazards by one of five types of controls: Elimination, substitution, engineering, administrative, and personal protective equipment (PPE). The term *hierarchy* seems to connote a process whereby an employer tries to pick one method of controlling a hazard, preferably a control more desirable on the hierarchy. The hierarchy of controls, in best practice, is actually a process through which employers begin by evaluating their ability to deploy the most desirable control, and then continue through the process, attempting to select the most desirable control. This process continues until the employer is left with no choice but to enact the least desirable controls. Figure 1.1 captures the hierarchy as a process.

Most desirable   Least desirable

FIGURE 1.1

Elimination is the preferred control as it means the hazard has been eliminated from the process. Elimination may include the discontinuation of an operation (akin to the insurance term of risk/peril avoidance) or the complete redesign of a manufacturing process that completely removes the worker from the hazardous activity.

Substitution is the next most preferred method of controlling a hazard in which a more serious hazard is replaced by a less serious hazard or no hazard. An example of substitution may include using a less hazardous cleaner in place of the prior, more hazardous substance.

Engineering controls are still a preferred method of control as this method seeks to remove the human element from the hazard, yet the hazard is still present. Examples of engineering controls may include the provision of lift assist devices or installation of barrier guards. A barrier guard, if properly designed and maintained, will prevent the worker from being exposed to the hazard. A lift assist, if properly designed and used by the employee, should remove the worker from exposure to the ergonomic hazard. Engineering controls are less desirable than elimination and substitution because the original hazard is still present, but they are still preferred over administrative controls and PPE.

Administrative controls involve taking steps to limit the worker's exposure to a hazard or educate the worker to manage the hazards when encountered. Examples of administrative controls include training, work rotation, checklists, permits, and so on. As with engineering controls, the original hazard is still present, but an administrative control's efficacy is dependent upon the human worker following training, work procedures, checklists, and so on. The reality is, though, that humanity cannot be detached from the work environment, and inevitably a worker may deviate, forget, or simply refuse to use the administrative control. If the employee's only protection is a procedure he or she may forget or elect not to follow, then the hazard exposure is not mitigated.

The least-desirable hazard control is PPE. PPE is least desirable because the exposure to the hazard remains and the only defense that lies between the worker and the hazard is the material from which the PPE is made. Similar to administrative controls, PPE also depends upon the worker to use the control, and human behavior can limit its effectiveness.

It is important to remember that there is no edict or absolute law that mandates only one type of hazard control can be used to address any one hazard. If the hazard is truly eliminated, coupling PPE with elimination is illogical, but it is possible and quite common for PPE to be used in conjunction with another hazard control. Safety professionals are encouraged to be innovative when prescribing controls and look for combinations that minimize employee exposure to hazards and risk.

HAZARD IDENTIFICATION AND CONTROL SYSTEMS

Performing hazard identification and control activities can, and frequently do, occur on an ad hoc basis. In most cases, ad hoc hazard identification and control activities are better than none at all. But performing hazard identification and control activities within a well-designed system is more desirable, and it's critical to an organization's overall occupational safety and health performance. While only a voluntary guideline, OSHA's Recommended Practices for Safety and Health Programs states: "A critical element of any effective safety and health program is a proactive, ongoing process to identify and assess … hazards." Furthermore, a system that sets forth the basic means for hazard identification (or at least reporting) and management of the process to control hazards are requirements of the leading, modern occupational safety and health management system standards (i.e., ANSI Z10 and ISO 45001). Some OSHA standards, such as the general industry standards on process safety management (29 CFR § 1910.119), permit required confined space entry (29 CFR § 1910.146), and general requirements for PPE (29 CFR § 1910.132) require varying degrees of formality with respect to the process of identifying and controlling hazards, and documentation of such. Accordingly, organizations are encouraged to develop a formal system that establishes processes for identifying and controlling hazards, and then evaluates the efficacy of the prescribed controls over time.

A survey of the OS&H management consensus standards, OSHA's Recommended Practices for Safety and Health Programs, and OSHA regulations will reveal a variation in the critical components and characteristics of a hazard identification and control system. All sources present the components in a slightly different fashion. Nonetheless, there is a common theme among these sources and the author's experience, and when combined and synthesized, the critical elements can be summarized as follows in ten elements:

1. *Employee Involvement.* Employee involvement must be a purposeful effort that involves more than a meeting where employees get to talk and eat donuts or pizza. A strategic plan must be developed at the outset that allows employees to help design tools and processes, as well as engage in the actual hazard identification and control activities.

2. *Management Commitment.* Equally important is management's strategic commitment to a robust hazard identification system and accountability for the proper assessment and control of hazards once identified. In order for the hazard identification and control system to be effective, management must be engaged in the identification and assessment process. More importantly though, management must lend support to control efforts and hold stakeholders accountable for both the timely and effective prescription and implementation of controls.

3. *Means for Workers to Report Hazards.* Whether it be a hazard report form, hazard hunt card, near miss report, mobile device app, or company intranet database, a means through which employees feel free to report hazards they detect in the workplace is crucial. Some hazards may not be apparent during an inspection or job safety analysis, even when done by the most seasoned safety professional. The employee who interacts daily with the processes, equipment, work conditions, procedures, and so on, may, however, realize a hazard during their work performance—often through a near-miss incident. When this occurs, a means by which the employee can memorialize the otherwise undetected hazard and make sure it is placed in the system queue for proper assessment and necessary control is imperative. Employers must

know that it is critical for supervisors, managers, safety staff, and so on, to recognize receipt of the reported hazard, thank the employee for submitting the report, and then continue to follow up with the employee on the final disposition of the hazard.

4. *Inspections and Audits.* Workplace inspections—activities that evaluate conditions and behaviors at a particular moment in time—are a critical part of an OS&H management system. Oftentimes the employer has created a document to guide the inspectors to evaluate the compliance status of controls prescribed for already-identified hazards. While this is indeed a hazard identification and control activity, the inspection process itself should not be limited to merely what is placed on a pre-printed inspection form. Inspectors should be encouraged to look for new hazards and there must be provisions to report on newly identified hazards in the inspection report document/system. Different from inspections, audits are usually larger in scope and look to evaluate an organization's compliance with established policies and procedures as well as regulatory requirements. During the audit process, documents are reviewed, workers are interviewed, and workplaces are inspected. The audit reporting process is normally designed to report on all findings and then ensure corrective action occurs.

5. *Hazard Reviews.* Intentional hazard reviews of processes, equipment, procedures, and so on, may take place at different phases of the operation's life cycle. Optimally, hazard reviews occur during the concept or design phase. Often these reviews will be required as part of an organization's safety through design (or safety by design) policy. Identifying hazards before equipment or processes are built or installed is the most cost-effective way to control a hazard as it prevents employee exposure and avoids the high cost of retrofit.

Other intentional hazard reviews may occur after a process or piece of equipment is already operational. More common in the general industry setting is the job safety analysis (JSA). JSAs involve breaking a job into distinct tasks, identifying the hazards associated with each task, and then either documenting or recommending required controls. Many organizations have integrated a risk assessment component in their JSA forms. Job hazard analysis (JHA) is often times used synonymously with JSA; however, there are many that make distinctions in the two. Purists would hold that JHAs are broader and frequently involve tools such as fault tree analysis (FTA) or failure modes and effects analysis (FMEA) during the review. Still, the hazard reviews seen in a construction context, such pre-job safety analysis and pretask safety analysis, are used to identify hazards and prescribe controls at a distinct, progressive phase of a bigger project. Regardless of the context, intentional hazard analysis activities must be prescribed by the hazard identification and control system. The system must establish what initiates the hazard reviews, the time frames in which the hazard reviews must be completed, communication of the completed hazard reviews, and any follow up necessary. (See Appendix A for an outline of the critical elements of a JSA System.)

6. *Hazard Assessment.* The hazard and risk assessment process has already been discussed at length in this chapter. Still, it is important to mention that assessment methods and expectations should be carefully set forth in the hazard identification and control system, along with guidance on what action is required when risk scores exceed certain thresholds.

7. *Control of Hazards.* As with hazard assessment, hazard control is discussed in a prior section of this chapter. Still, it is important for organizations to realize that hazard identification and control policies and procedures should explicitly call for the application of more desirable controls such as elimination, substitution, or engineering controls before accepting less-desirable controls like PPE. Some organizations even require documentation to show how higher-order controls were considered and deemed infeasible whenever lower-order, less-desirable controls are prescribed. Minimally, those involved in the hazard control process must be trained to always first look for higher-order controls.

8. *Communication of Hazards.* The hazard identification and control system should ensure that mechanisms are in place to ensure that requirements for proper hazard controls are quickly communicated through job instructions, training, and so on. In cases where only a short-term control for a newly identified hazard has been prescribed and safety teams are still determining the most desirable long-term methods, affected workers should be notified of the hazard, the risks it poses, and most importantly the necessary, interim protective measures that should be taken. Frequently companies create some version of a hazard bulletin to serve this quick information need. Hazard bulletins are particularly helpful when sharing information among multiple locations that have the same hazard.

9. *Evaluation of Control Effectiveness.* After a hazard control is implemented, an organization should require an intentional, data-driven effort to analyze the effectiveness of that hazard control. The simple absence of further injury is not necessarily indicia of a hazard control's effectiveness. Post-control implementation, organizations should observe behaviors, take measurements, survey employees, and so on, to obtain data that can be analyzed through statistical process control or used in calculating correlations. The absence of injuries or near-misses should be viewed simultaneously with these types of calculations, and both should support any conclusion that a control is effective. For example, if the prescribed hazard control is training, then post-training observations should reveal a sustained increase in the desired behavior/activity that prevents injury. If after training there have been no injuries, yet observation data does not reveal an improvement in behavior, the training control is not effective, and the absence of injury may simply be attributed to good luck.

10. *Evaluation of System Effectiveness.* Finally, and just as important as evaluating the efficacy of hazard controls is the ongoing evaluation of the hazard identification and control system itself. Often the evaluation of the system's effectiveness, and the organization's proper usage of the system, will be a key component of a comprehensive safety and health management system

audit. In essence, the key objectives for evaluating the hazard identification and control system is to ensure the system's design is still appropriate for the organization or that the organization is properly following the system as established. Any negative finding with either of these objectives should warrant corrective action.

SUMMARY

The fundamental tasks of the safety profession involve the anticipation, identification, assessment, and control of hazards. The terms *hazard* and *risk* are distinctly different, and it is important to understand that a hazard is the source of harm while a risk represents some quantification of threat that harm presents. Safety professionals should strategically develop systems through which hazard identification and control occurs in order to maximize chances of success. After hazards are identified and assessed, appropriate controls must be prescribed and the ongoing effectiveness of these controls must be evaluated.

Review Questions

1. Name the fundamental, essential functions of the safety profession.
2. Define and distinguish the terms *hazard* and *risk*.
3. In your own words, describe what is meant by a *physical hazard*.
4. In your own words, explain why the term *safety hazard* is arguably an oxymoron.
5. Explain the most important, contextual considerations for four of the means of hazard identification mentioned in this chapter.
6. What is one key, legal consideration when documenting identified hazards?
7. Why has there been a trend to expand the number of variables used when assessing/calculating risk?
8. In your own words, and in less than three sentences, explain why PPE and administrative controls are the least desirable methods to control hazards.
9. Name one, specific example of an engineering control you've personally encountered.
10. How should the effectiveness of a hazard control be evaluated after it is implemented?

REFERENCE

Occupational Safety and Health Administration. 2016. *Recommended Practices for Safety and Health Programs*, OSHA Publication 3885. Washington, DC. Accessed from www.osha.gov/shpguidelines.

2 Ergonomics

Just before the first edition of this book was published, OSHA issued one of the most contentious final rules in its history: The Ergonomics Program Rule. This broadly applicable ergonomics rule was enacted at the end of the Clinton Administration and had been anticipated for some time. While ergonomics was—and continues to be— the hazard arguably responsible for most lost time injuries, restricted work cases, and workers' compensation costs, the need for and best way to regulate ergonomics was the subject of much debate and disagreement. After the publication of the first edition and the inauguration of President George W. Bush, Congress for the first time used the Congressional Review Act (CRA) to repeal a regulation promulgated by an executive branch agency. The effect of this CRA repeal of the Ergonomics Program Rule was twofold: (1) it eliminated OSHA's Ergonomics Program Rule, and (2) it prevents OSHA from ever issuing another regulation that is substantially the same unless Congress, through legislative action, specifically authorizes OSHA to do so. In essence, it will take an act of Congress in order for another comprehensive ergonomics regulation to ever be created.

Since the CRA repeal of the Ergonomics Program Rule, OSHA has worked diligently to produce industry-specific guidelines to direct those particular employers on best practices in reducing ergonomic injuries. Also, OSHA has continued to issue citations for ergonomic hazards in the same manner it did prior to enactment of the Ergonomics Program Rule—through Section 5(a)(1) of the OSH Act, the General Duty Clause. Accordingly, employers must be aware that they still face OSHA liability for ergonomic hazards in the workplace.

Readers should also consider that long before OSHA ever enacted the Ergonomics Program Rule, companies had already created their own ergonomics programs to improve worker health and control skyrocketing workers' compensation costs. The vast majority of even quasi-progressive companies have understood the benefits of sound ergonomic initiatives, and have invested significant resources to reduce ergonomic-type injuries. Accordingly, the reader should examine the timeless content of this chapter when implementing an ergonomics program from scratch or honing an already existing system.

Ergonomics is defined as the science of relating the worker to all aspects of the job and the job environment. The word *ergonomics* breaks down into two words: *Ergo* means the act of work, while *nomics* means the law of work; therefore, ergonomics is "the law of work." Several terms are associated with ergonomics, including the following:

- *Cumulative trauma disorder* (CTD): Disorder that arises from repeated stress and develops as a result of chronic exposure of a particular body part to repeated stress; for example, use of the same, repetitive knifing motion by a meat cutter while boning could result in a CTD.

- *Carpal tunnel syndrome*: Compression of the medial nerve in the carpal tunnel, a passage in the wrist through which finger tendons and a major nerve pass to the hand from the forearm. It is often associated with tingling, pain, or numbness in the thumb and first three fingers.
- *Tendinitis*: Inflammation of the muscle–tendon junction and adjacent muscle tissues resulting from repeated stress of a body member.
- *Tennis elbow*: Inflammation of tissue in the elbow.
- *Trigger finger*: Condition in which the finger frequently flexes against resistance.

In sum, ergonomic hazard controls are an attempt to address the multitude of cumulative trauma and overexertion injuries and illnesses, such as carpal tunnel syndrome, which have inflicted substantial pain and disability upon employees performing various industrial and administrative tasks. Addressing ergonomic hazards usually involves the deployment of all of the hazard controls from the hierarchy. Most prevalent among the viable ergonomic hazard controls are engineering controls, administrative controls, and PPE. Engineering controls may include the design and provision of lift-assist devices, adjustable work stations, and work station redesign. Administrative controls may include work rotation schedules for repetitive tasks to reduce employee exposure, preventive stretching and exercise programs, training on proper posture/ lifting techniques, and medical management activities. Personal protective equipment may include items such as anti-vibration gloves. It is important to note, however, that these controls are usually prescribed within a well-designed ergonomics management program. The proceeding content provides key considerations and guidance for the development and implementation of an effective ergonomics program.

DEVELOPING AND IMPLEMENTING THE ERGONOMICS PROGRAM

MANAGEMENT COMMITMENT AND EMPLOYEE INVOLVEMENT

The first step in developing an ergonomic program is to gain management commitment. This will provide the organizational resources and motivating force necessary to deal effectively with ergonomic hazards. Employee involvement and feedback are likewise essential for identification of existing and potential hazards and for development and implementation of this program.

To better facilitate employee involvement in the ergonomic process, the employer should consider creating and developing teams for identifying and correcting ergonomic hazards in the workplace. The team should consist of a wide range of personnel in the facility. Members should include such personnel as members of the safety department, occupational health/nursing department, production managers, production supervisors, and maintenance personnel.

WORKSITE ANALYSIS

The next important step of the ergonomic process is to perform a worksite analysis. This analysis may be performed by the ergonomic team and involves examining and identifying existing hazards or conditions and operations that may create a hazard. During the analysis, it is recommended that medical, safety, and workers' compensation records be evaluated for evidence of cumulative trauma disorders. The worksite analysis should identify those work positions requiring an analysis of ergonomic hazards. This analysis should encompass the following:

- Use of an ergonomic checklist that includes components such as posture, force, repetition, vibration, and various upper extremity factors.
- Identification of work positions that put workers at risk of developing cumulative trauma disorders.
- Verification of low-risk factors for light-duty jobs and restricting awkward work positions.
- Verification of risk factors for work positions which have already been evaluated and corrected to the extent feasible.
- Providing results of worksite analysis for assigning light-duty jobs.
- Re-evaluation of all planned, new, and modified facilities, processes, materials, and equipment to ensure that workplace alterations contribute to reducing or eliminating ergonomic hazards.

It is recommended that periodic surveys be conducted at least annually or when operations change or the need arises. They should be conducted to identify new deficiencies in work practices or engineering controls and to assess the effects of changes in the work processes.

In the 2001 Ergonomics Program Rule that was repealed by the CRA, OSHA had included a rather exhaustive outline of ergonomic risk factors to consider when performing a worksite analysis. The outline is still instructive for safety professionals and ergonomic teams. The outline is as follows:

Ergonomic risk factors associated with a physical work act or circumstances surrounding a work activity.
1. Exerting considerable physical effort to complete a motion
   i. Force
   ii. Awkward postures
   iii. Contact stress
2. Doing the same motion over and over again
   i. Repetition
   ii. Force
   iii. Awkward postures
   iv. Cold temperatures
3. Performing motions constantly without short pauses or breaks in between
   i. Repetition
   ii. Force
   iii. Awkward postures
   iv. Static postures
   v. Contact stress
   vi. Vibration

4. Performing tasks that involve long reaches
   i. Awkward postures
   ii. Static postures
   iii. Force
5. Working surfaces that are too high or too low
   i. Awkward postures
   ii. Static postures
   iii. Force
   iv. Contact stress
6. Maintaining same position or posture while performing tasks
   i. Awkward posture
   ii. Static postures
   iii. Force
   iv. Cold temperatures
7. Sitting for a long time
   i. Awkward posture
   ii. Static postures
   iii. Contact stress
8. Using hand and power tools
   i. Force
   ii. Awkward postures
   iii. Static postures
   iv. Contact stress
   v. Vibration
   vi. Cold temperatures
9. Vibrating working surfaces, machinery, or vehicles
   i. Vibration
   ii. Force
   iii. Cold temperatures
10. Workstation edges or objects that press hard into muscles or tendons
   i. Contact stress
11. Using hand as a hammer
   i. Contact stress
   ii. Force
12. Using hands or body as clamp to hold object while performing tasks
   i. Force
   ii. Static postures
   iii. Awkward postures
   iv. Contact stress
13. Gloves that are bulky, too large, or too small
   i. Force
   ii. Contact stress

Ergonomic risk factors that may be present for manual handling (lifting/lowering, pushing/pulling and carrying)

14. Objects or people moved that are heavy
     i.   Force
     ii. Repetition
     iii. Awkward postures
     iv. Static postures
     v.   Contact stress

15. Horizontal reach that is long (distance of hands from body to grasp object to be handled)
     i.   Force
     ii. Repetition
     iii. Awkward postures
     iv. Static postures
     v.   Contact stress

16. Vertical reach that is below knees or above the shoulders (distance of hands above the ground when the object is grasped or released)
     i.   Force
     ii. Repetition
     iii. Awkward postures
     iv. Static postures
     v.   Contact stress

17. Objects or people that are moved significant distance
     i.   Force
     ii. Repetition
     iii. Awkward postures
     iv. Static postures
     v.   Contact stress

18. Bending or twisting that occurs during manual handling
     i.   Force
     ii. Repetition
     iii. Awkward postures
     iv. Static postures

19. Object that is slippery or has no handles
     i.   Force
     ii. Repetition
     iii. Awkward postures
     iv. Static postures

20. Floor surfaces that are uneven, slippery, or sloped
     i.   Force
     ii. Repetition
     iii. Awkward postures
     iv. Static postures

Hazard Prevention and Control

Once ergonomic hazards are identified through worksite analysis, the next step is to design measures to prevent or control these hazards. Ergonomic hazards are primarily prevented by effective design of the workstation, tools, and the job.

1. Engineering Controls

   Engineering techniques are the preferred method of correcting ergonomic hazards. The purpose of engineering controls is to make the job fit the person, not the person fit the job. This can be accomplished by designing or modifying the workstation, work methods, and tools to eliminate excessive exertion and awkwardness.

   • Workstation Design

   A workstation should be designed to accommodate the person who actually works on a given job; it is not adequate to design for the average or typical worker. Workstations should be easily adjustable and designed or selected to fit specific tasks so that they are comfortable for the workers using them. The work space should be large enough to allow the full range of required movements, especially where knives, saws, hooks, and similar tools are used.

   • Design of Work Methods

   Work methods should be designed to reduce static, extreme, and awkward postures; repetitive motions; and excessive force. Work method design addresses the content of tasks performed by the workers. It requires analysis of the production system to design or modify tasks to eliminate stressors.

   • Tool and Handle Design

   Tools and handles, if well designed, reduce cumulative trauma disorders. For any tool, a variety of grip sizes should be available to achieve a proper fit and reduce ergonomic risk. The appropriate tool should be used to do a specific job. Tools and handles should be selected to eliminate or minimize the following stressors:
   – Chronic muscle contraction or steady force
   – Extreme or awkward finger/hand/arm positions
   – Repetitive forceful motions
   – Tool vibration
   – Excessive gripping, pinching, pressing with the hand and fingers

2. Administrative Controls

   Administrative controls are those that reduce the duration, frequency, and severity of exposures to ergonomic stressors. Those controls include:

   • *Work practice controls.* These practices include safe and proper work habits that are understood and followed by managers, supervisors, and workers. The elements include proper work techniques, employee conditioning, regular monitoring, feedback, maintenance, adjustments, modifications, and enforcement.

- *Proper work techniques.* These techniques should include appropriate training and practice time for employees, such as
    - Proper cutting techniques, including work methods that improve posture and reduce stress and strain on extremities
    - Good knife techniques, including steeling and the regular sharpening of knives
    - Correct lifting techniques, including proper body mechanics (e.g., using the legs while lifting, not the back)
    - Proper use and maintenance of pneumatic and other power tools
    - Correct use of ergonomically designed workstations and fixtures
- *New employee conditioning period.* New and returning employees should be gradually integrated into a full workload. Employees should be assigned to an experienced trainer for job training and evaluation during this break-in period.
- *Monitoring*: Regular monitoring of the workplace should ensure that employees are continuing to use proper work practices. This monitoring will include a periodic review of techniques and a determination of whether procedures in use are those specified; if not, then it should be determined why changes have occurred and whether corrective action is necessary.
- *Adjustments and modifications*: These are essential when changes occur in the workplace. Such adjustments could apply to the following:
    - Line speeds
    - Staffing of positions
    - Type, size, weight, or temperature of the product handled
    - Reducing the total number of repetitions per employee by means of decreasing production rates and limiting overtime work
    - Providing rest pauses to relieve fatigued muscle–tendon groups
    - Increasing the number of employees assigned to a task, thus alleviating severe conditions, especially while lifting heavy objects
    - Job rotation as a preventative measure (the principle of job rotation is to alleviate physical fatigue and stress of a particular set of muscles and tendons by rotating employees among other jobs that use different motions; the ergonomic team should review a job analysis of each operation to ensure that the same muscle–tendon groups are not used)
    - Providing sufficient standby and relief personnel for a foreseeable condition on production lines
    - Preventative maintenance for mechanical and power tools and equipment, including power saws and knives
    - Knife-sharpening program (sharp knives should be readily available)
    - Personal protective equipment
- Medical Management

    Implementation of a medical management system is a major component of an ergonomic program. It is necessary to eliminate and reduce

the risk of development of cumulative trauma disorder signs and symptoms through early identification and treatment. The medical management system should address the following:
- Injury and illness recordkeeping
- Early recognition and reporting
- Systematic evaluation and referral
- Conservative treatment
- Conservative return to work
- Systematic monitoring
- Adequate staffing and facilities

If feasible and warranted, members of a medical management team should conduct workplace walkthroughs to remain knowledgeable about operations and work practices to identify potential light-duty jobs and to maintain close contact with employees. A survey of employees should be conducted to measure employees' awareness of work-related disorders and to report the location, frequency, and duration of discomfort. Not including employee names on surveys encourages employee participation in the survey. The major strength of the survey is collection of data on the number of workers that may be experiencing cumulative trauma disorders. The survey should be conducted annually to help determine any major changes in severity, incidence, or location of reported symptoms.

- Training and education

The purpose of training and education is to ensure that employees are sufficiently informed about ergonomic hazards to which they may be exposed and thus are able to participate actively in their own protection. Training and education should provide an overview of potential risks of illness and injuries, their causes and early symptoms, the means of prevention, and treatment. Training allows managers, supervisors, and employees to understand ergonomics and other hazards associated with the job or the production process, their prevention and control, and their medical consequences. A training program should include the following:
- All affected employees
- Engineers and maintenance personnel
- Supervisors
- Managers
- Medical personnel
  General training

  Employees who are potentially exposed to ergonomic hazards should be given formal instruction on the hazards associated with their jobs and with their equipment. This includes information on the varieties of cumulative trauma disorders, what risk factors cause or contribute to them, how to recognize and report symptoms, and how to prevent these disorders.

Job-specific training

New employees and reassigned workers should receive an initial orientation and hands-on training prior to being placed in a full-time production job. Training production lines may be used for this purpose. Each new hire should receive a demonstration of the proper use of and procedures for all tools and equipment. The initial training program should include the following:

- Care, use, and handling techniques of tools
- Use of special tools and devices associated with individual workstations
- Use of appropriate lifting techniques and devices
- Use of appropriate guards and safety equipment, including personal protective equipment

Training for supervisors

Supervisors are responsible for ensuring that employees follow safe work practices. Supervisors should undergo training comparable to that of the employees in addition to further training that will enable them to recognize early signs and symptoms of cumulative trauma disorders, to recognize hazardous work practices, and to correct such practices.

Training for managers

Managers should receive training pertaining to ergonomic issues at each workstation and in the production process so they can effectively carry out their responsibilities.

Training for maintenance

Maintenance personnel should be trained in prevention and correction of ergonomic hazards through job and workstation design and proper maintenance.

- Auditing

The ergonomic team should review the ergonomic program to ensure that new and existing ergonomic hazards are identified and corrected. If the need arises, corrective measures can be taken to eliminate or minimize the hazards.

3. Personal Protective Equipment

Personal protective equipment (PPE) should be selected with ergonomic stressors in mind. PPE should be provided in a variety of sizes, should accommodate the physical requirements of workers and the job, and should not contribute to ergonomic hazards. The following PPE should be considered for purchase:

- Gloves that facilitate the grasping of tools required for a particular job while protecting the worker from injury
- Apparel that provides protection against extreme temperatures
- Braces, splints, and back belts for support

ERGONOMIC PROGRAM CHECKLIST

The following checklist is presented to assist the reader in assessing progress in developing and implementing an ergonomics program at her or his facility:

- Has an ergonomic team been developed in the facility?
- Has a worksite analysis been conducted to identify ergonomic hazards?
- Have posture, force, repetition, vibration, and various other factors been observed? If so, has equipment been modified or has PPE been issued to eliminate ergonomic hazards?
- Have periodic surveys been conducted to check for new and existing ergonomic hazards?
- If ergonomic hazards have been identified, have measures to prevent or control these hazards been developed?
- In all operations, does the job fit the person or does the person fit the job?
- Are all work spaces large enough to allow for the full range of required movements?
- Are work methods designed to reduce static, extreme, and awkward postures?
- Has an analysis of the production process been conducted to identify and eliminate ergonomic stressors?
- Are new and returning employees gradually integrated into a full workload?
- Is the workplace monitored to ensure employees continue to use proper work procedures?
- Has overtime work been decreased to limit the total number of repetitions per employee?
- Are employees given breaks to relieve fatigued muscle–tendon groups?
- Has the job rotation principle been implemented to alleviate physical fatigue and stress of a set of muscle–tendon groups?
- Has each job been reviewed by the ergonomic team to identify ergonomic hazards?
- Are sufficient standby personnel available for unforeseeable conditions on production lines?
- Is preventative maintenance performed on tools and equipment to limit the amount of force an employee has to exert to perform a task?
- Is PPE available to employees in a variety of sizes?
- Has a medical management system been implemented to research workers' compensation costs for evidence of CTDs?
- Are employees trained and educated regarding ergonomic hazards that they may encounter?
- Does training provide an overview of potential risks of illnesses and injuries of ergonomic hazards?
- Is specialized training conducted specifically for management, supervisors, and maintenance personnel?
- Is the ergonomic program audited annually?

- Has your ergonomic policy been posted for employee observation?
- Have employees been advised of your facility's disciplinary policy for noncompliance?

SUMMARY

Cumulative trauma disorders and other ergonomic hazards represent a substantial cost to employers, and the injuries and illnesses cause employees remarkable pain, loss of life activities, and other harms. Safety professionals should address the potential ergonomic risk factors in their workplace and take appropriate measures to eliminate or minimize these risks. Development of an effective ergonomics program has become a "must" for most employers, and the dividends realized from an effective ergonomics program are substantial for both employees and the employer.

Review Questions

1. Explain OSHA's current enforcement mechanism for ergonomic hazards in the workplace.
2. Explain what needs to happen in order for OSHA to promulgate an ergonomics rule.
3. Distinguish carpal tunnel syndrome from regular tendinitis.
4. Survey the work acts/conditions/work circumstances and associated ergonomic risk factors, and then create a simple yet exhaustive list of all the different ergonomic risk factors.
5. What are some key provisions and considerations of the medical management component of an ergonomics program?

3 Respiratory Hazards

Even though respiratory hazards frequently involve chemicals—hazards not typically thought of as physical hazards—the evaluation and control of respiratory hazards is a pervasive issue facing safety practitioners in a wide variety of industries. Furthermore, the safety practitioner is frequently called upon to identify and control respiratory exposures even when a permissible exposure limit is not exceeded and no appreciable risk to health is present. For example, certain odors and nuisance dust can make work uncomfortable even though it doesn't threaten an employee's health. Finally, respiratory hazards many times involves solid particles (dust) that may be harmful to the human body not because of the chemical composition, but for the mere fact that the solid particle is small enough to be inhaled. For these reasons, a discussion of respiratory hazards is appropriate even in a work dedicated to physical hazards.

Failure to respect respiratory hazards can kill workers in a matter of seconds. Poor work practices can also allow toxic materials to slowly accumulate, and their effects may surface many years later in the form of chronic lung illness, cancers, or diseases of specific target organs. It is imperative that all employees fully understand the physiological aspects of the respiratory system, limitations of the various personal protective equipment (PPE) available, and responses of the body to respiratory hazards. If employees understand their bodies, the limitations of the personal protective equipment, and their personal limitations, they can make intelligent and safe decisions to protect themselves.

Improperly compensated risk is a primary factor in many unnecessary respiratory hazard incidents. The employees do not have enough information about the material they are being exposed to or they may not understand the limitations of their PPE. Although experience can be a good teacher, it can also reinforce negative behaviors. An employee who does not wear a respirator and completes the task with no apparent ill effects from the exposure to toxic gases, vapors, or airborne particulate hazards may be more likely to repeat the dangerous practice.

At a firefighter safety and health symposium, a cancer researcher described common carcinogens firefighters are exposed to in the line of duty. After learning about the dangers of asbestos, a group of firefighters described a daily practice in their station of removing an asbestos fire blanket from an apparatus compartment to gain access to an engine-powered rotary saw. After removing the blanket and rescue saw, they would place the saw on the ground near the blanket and start the saw to ensure that it would run if needed on their tour of duty. This procedure was performed every morning and again by the evening shift when they relieved the day shift. The researcher shuddered as they talked. Each time the blanket was handled, friable pieces of the asbestos blanket were undoubtedly suspended in the air in the fire house. When the saw was started in close proximity to the blanket, additional fibers were blown all over the station by the engine exhaust. The firefighters did not have a death wish. They just failed to realize the risk they were taking. It is the employer's responsibility under 29 CFR § 1910.134 to make sure everyone in its facilities understands all the respiratory hazards and follows safe work practices.

All employees may not be physically fit to wear certain respirators. It is the employer's responsibility to provide medical screening for all employees who may have to wear respirators. OSHA has developed a simple questionnaire that can be administered to determine if a particular employee must be screened by a physician. A copy of that questionnaire is included in Appendix D.

RESPIRATORY HAZARD EXPOSURE PREVENTION METHODS

Policies, procedures, signs, and permit systems are examples of administrative control methods used to prevent employees from entering certain areas where hazards exist. These same methods may be employed to allow entry into certain areas only after the area(s) have been determined to be safe; also, other means to protect the worker have been approved and documented. A confined-space entry permit system is an example of an administrative control designed to prevent worker exposure and injury.

A common engineering control used for controlling respiratory hazards is a ventilation hood system or local fume mitigation vacuum. A fan system pulls toxic vapors, fumes, or dust away from the employee workstation and exhausts them in a safe area.

In controlled industrial hazard settings, PPE is usually the last line of defense. If the hazard cannot be isolated using administrative controls or engineering controls, then PPE is selected specifically for the type of hazard. If the concentration of the hazard present in the air can be measured, an appropriate respirator with a suitable protection factor is selected to ensure that employees are not exposed above acceptable levels. If the exact hazard is unknown, if multiple hazards exist, or if the substance is highly toxic and air-purifying respirators will not provide adequate safety, air-supplied respirators are required. Firefighters, hazardous materials teams, and other first responders rarely have the ability to measure the quantity of substance(s) in the air, so their first choice is usually the positive-pressure, self-contained breathing apparatus (SCBA). These full-facepiece respirators have high-pressure cylinders that supply breathing gas independent of the hazardous atmosphere. They offer the highest level of protection available. Quantitative fit testing should be performed to ensure that these individuals are able to wear the SCBA without leaks around the sealing edge of the facepiece. Most respirator manufacturers offer facepieces in several sizes to accommodate a wide variety of face shapes.

CLASSIFICATION OF RESPIRATORY HAZARDS: GASES AND FUMES

Simple asphyxiants are gases that displace the air that contains the oxygen required for life. All of the inert gases, such as carbon dioxide, argon, nitrogen, and helium, are examples of simple asphyxiants. Gases with a vapor density heavier than air (greater than 1.0) pose potential hazards to workers in basements and other areas such as confined spaces. Gases with a vapor density lighter than air (less than 1.0) can accumulate at the top of shafts or ceilings. One of the major problems associated with these simple asphyxiants is the fact that they may be odorless, colorless, and tasteless. A worker can enter a space filled with these gases and collapse after inhaling a single breath with no oxygen.

Toxic gases are those that cause problems in very low concentrations. Carbon monoxide is considered immediately dangerous to life and health at concentrations of 200 parts per million. If the particular gas or vapor is also flammable, you need to determine the upper and lower flammable limits. Methane, for example, has a lower flammable limit of 5 percent and an upper flammable limit of 15 percent. This means that outside of that narrow 10 percent range, the mixture will either be too lean (below 5 percent) or too rich (above 15 percent) to burn. Some people erroneously believe it is safe to work in concentrations too rich to ignite. This is a dangerous, even deadly, misconception.

The bottom line is, do not enter areas containing more than 10–20 percent of the lower flammable limit of the gas being measured. Also, make sure you either calibrate your combustible gas instrument with the gas being tested or use a chart furnished with the instrument to interpret the results on the challenge gas based upon your calibration gas. When selecting combustible gas instruments, choose continuous or intermittent monitors over spot analyzers, which only provide color change or visual meter indications. Continuous or intermittent monitoring units sample and interpret the atmosphere when workers in the area are busy and will sound an alarm and flash lights to warn the incident commander, thus allowing personnel time to egress from the space before ignition.

Particulates are particles of solid material suspended in the air. These can be inhaled and trapped in the respiratory system. The body has several lines of defense against these, but their ability is limited and should not be relied upon to protect the worker. Almost any type of dust can be carried deep into the lungs and cause irreversible damage to delicate airways and alveoli. Chronic obstructed pulmonary disease (COPD) is treatable if diagnosed early and the exposure is stopped. "Black lung," which is a form of COPD, has ruined the lives of many coal miners who were exposed to coal dust. Other workers in many industrial settings are exposed to other dusts that will yield the same result. In advanced stages of COPD, the patient's lungs are unable to exchange oxygen and carbon dioxide. Eventually, the patient's lungs begin to fill with fluid; some eyewitness accounts describe the patient's death as one of drowning in fluid trapped in the patient's lungs. Smoke from tobacco products such as cigarettes causes the lungs to lose their elasticity and compounds the problems of this disease. Employer dollars spent on stopping smoking will come back to the employer in increased worker productivity, less absenteeism, savings on healthcare costs, and improved quality of life for workers. Several fire departments in the U.S. require new hires to sign agreements indicating they will not use tobacco products.

Another common industrial particulate is silica. More than 1 million U.S. workers are exposed to crystalline silica. Each year, more than 250 American workers die from silicosis. There is no cure for the disease, but it is 100 percent preventable if employers, workers, and health professionals work together to reduce exposures. The smaller the silica particle size, the deeper into the human airway it travels. The OSHA standard specifically addresses this by limiting the total allowable silica dust and respirable silica dust. In 2016, OSHA issued revised rules on respirable silica for construction, general industry, and maritime workplaces. These revised rules are contained in Appendices E and F.

Asbestosis is another deadly airborne particulate. An estimated 1.3 million employees in construction and general industry face significant asbestos exposure on the job. Heaviest exposures occur in the construction industry, particularly during removal of asbestos during renovation or demolition. Asbestos exposure is highly regulated by both OSHA and the Environmental Protection Agency (EPA). Asbestos fibers can have serious effects on a person's health if inhaled. There is no known safe exposure to asbestos. The greater the exposure, the greater the risk of developing an asbestos-related disease. The amount of time between initial exposure to asbestos and the first signs of disease can be as long as 30 years. Smokers exposed to asbestos have a higher risk of developing lung cancer than from just smoking alone. Workers who are exposed to asbestos products are often affected with asbestosis, a scarring of the lungs that leads to breathing problems and heart failure. Inhalation of asbestos fibers can cause mesothelioma, a rare cancer of the linings of the chest and abdomen. The American Lung Association indicates it may also be linked to cancers of the stomach, intestines, and rectum. Again, the good news is that these diseases can be prevented through following safe work practices and proper use of PPE.

Environmental considerations can impact the performance of the respirator and the person using it. In extremely cold conditions, the edges of the facepiece may become rigid and fail to maintain an adequate seal, allowing toxic gases to be inhaled. If the breathing air in the cylinders is not processed to Compressed Gas Association Breathing Air Standards of Grade D or better, any moisture in the air can freeze into ice bullets and cause regulators to fail. In extremely toxic environments this can be fatal. In high-temperature environments, the wearer's perspiration may allow the facepiece to slide and loosen the seal between the face and mask. The breathing gas in the SCBA cylinder remains the firefighter's only means for surviving in high-temperature environments where a single breath of superheated air would blister his fragile lungs. The hazardous materials technicians and specialists also work in environments where a single breath of a toxic atmosphere could kill them. This lifesaving air supply is not delivered to the wearer without exacting a price. SCBA rated with a 30-minute service life weighs more than 20 pounds, adding to the physical stress on the body. In situations requiring 60 minutes' or greater duration, the SCBA can weigh as much as 34 pounds. Whether a Level A encapsulating chemical suit or a firefighter's standard turnout clothing is worn, the wearer must be in excellent physical condition to work safely and effectively.

DIFFERENT TYPES OF RESPIRATORS AVAILABLE

The Occupational Safety and Health Administration has elected to recognize and approve only respirators approved by the National Institute for Occupational Safety and Health (NIOSH) or the Mine Safety and Health Administration (MSHA) as set forth in 42 C.F.R. Part 84.

Air-purifying respirators (APRs) are available in a wide range of styles but all of them attempt to filter toxins from the environment or absorb chemicals as the air passes through the filter medium. APRs are grouped into one of three classes: Particulate-removing APRs utilize filters to reduce inhaled concentrations of dusts, mists, fumes, or fibers. Vapor- and gas-removing APRs are designed with sorbent cartridges or canisters that absorb the vapors or fumes before they reach the wearer's airway. The third style of APR combines particulate and vapor/gas removal in a single unit.

Air-purifying respirators can be either nonpowered or powered air purifying, with blowers to pull the contaminated air through the filter/canister. APRs can be single use (disposable) or can utilize replaceable filters and/or canisters or cartridges. Some APRs are designed for specific hazards such as *M. tuberculosis* (29 C.F.R. 1910.139). Some chemical cartridge respirators are designed to remove specific gases, such as ammonia. All of these filter masks can only be worn by a user if the concentration of contaminant in the air is known and the APR has a high enough fit factor as demonstrated by qualitative fit testing (QLFT) or quantitative fit testing (QNFT) of that individual. The oxygen concentration in the room must also be equal to or greater than 19.5 percent.

Supplied-air respirators (SAR) have an advantage over APRs because the breathing gas is being supplied from an independent source rather than being filtered. If the SAR maintains the breathing gas at a slightly higher pressure than the ambient air, the SAR is considered positive pressure or pressure demand. This minimizes the opportunity for toxins to be inhaled during inspiration. SARs may not be used in atmospheres immediately dangerous to life and health without an escape air supply. This permits the worker to egress if the flow of breathing gas is interrupted for any reason. One of the most desirable combinations for long-duration use is a SAR used in conjunction with a 30-minute, pressure-demand SCBA. This enables the worker to spend a reasonable time in the contaminated area (hot zone) and still have ample time to decontaminate without fear of exhausting the breathing gas supply. The only disadvantage to this combination is the potential for tangling the supplied air hose and trapping the worker.

The self-contained breathing apparatus affords the highest level of protection. It consists of a high-pressure, DOT-regulated cylinder (typical cylinder pressures range from 2216 psig to 4500 psig). Cylinder capacities are 45 feet^3 for 30-minute units and approximately 90 feet^3 for 60-minute units. The high-pressure gas is reduced to very low pressure—2 to 3.5 in. of water column maximum—by a series of regulators designed to maintain this slight pressure and never allow the pressure in the facepiece to drop lower than the atmospheric pressure outside the facepiece. In this manner, if the facepiece leaks slightly, breathing gas leaks out rather than toxins leaking into the facepiece. All approved respirators employ an end-of-service-life warning device which must sound when 20–25 percent of the pressure remains in the cylinder. This is designed to allow the wearer to egress to a safe atmosphere. The units also feature pressure gauges that enable the wearer to check the pressure remaining at any time.

OPEN-CIRCUIT VS. CLOSED-CIRCUIT SCBA

Although most SCBA in use today is of the open-circuit type described in the preceding paragraph, another type of SCBA, the closed-circuit type (CCSCBA), has enjoyed considerable use for many years for underground mine rescue use. Like the open-circuit units, the CCSCBA employs a high-pressure cylinder of breathing gas. In the CCSCBA, however, the breathing gas is not Grade D breathing air; it is medical-grade oxygen. Instead of each breath coming straight from the high-pressure cylinder, the user's exhaled breath is passed through an absorbent canister filled with a material that absorbs the carbon dioxide from the exhaled air. The remaining oxygen flows into a breathing reservoir where it is available for the user's next breath. Only the amount of oxygen actually used by the body must be replaced from the cylinder. This allows a relatively small cylinder of breathing gas to last a comparatively long time. Mine-rescue teams have used two- and four-hour CCSCBA for decades, and the units weigh less than 34 pounds. The ability to produce positive-pressure CCSCBA emerged in the early 1980s, and today some of these units are approved for use in IDLH (immediately dangerous to life and health) atmospheres but not for use in structural firefighting. A disadvantage of the CCSCBA is that each time the gas cylinder is changed, the CO_2 absorber must be changed, and it must be stored in a sealed manner to protect the absorbent material from the CO_2 that exists in ambient air (approximately 0.05 percent). When cleaning the respirator, the breathing chamber and hoses must be cleaned and disinfected in addition to the facepiece. Even with all these limitations, for certain long-duration applications where a tethered air-line supply is impractical or dangerous, this unit may be the best choice. Another variation of the CCSCBA involves the use of liquid oxygen instead of a high-pressure cylinder. These units have also been used by mine-rescue teams.

RESPIRATOR TRAINING

Although 29 CFR § Part 1910.134(c), below, addresses specific training requirements, additional information is provided here to assist the program administrator/instructor. After determining that the individuals who will be trained are physically and psychologically fit to wear the respirator(s) required for the task, the instructor must take into account several factors when designing the training. First is the level of experience/exposure to respirators. If a person was able to hide a claustrophobic reaction to the facepiece during the initial physician screening, this reaction may still surface during the training. If these individuals are allowed to slowly become comfortable with the limited peripheral vision, they may be able to tolerate the respirator. A trainer who is fortunate enough to have individuals in the group who have experience and are comfortable with the respirators can spend more time working with those who are less than excited about wearing them.

There is no substitute for knowledge about the hazard, the respirator, and the limitations of the user and the respirator. People who understand the dangers of the hazardous agent and the sealing characteristics of the facepiece rarely try to wear beards, sideburns, and glasses with temple bars that impede the seal of the mask. Quantitative fit testing will resolve any questions about the ability to develop an effective seal with beards, and so on. Give these new users detailed information about the hazards and the long-term effects of exposure to the hazards. Make sure they understand the operation of the end-of-service-life indicator for both APR filters/canisters and SCBA service alarms. As with any aspect of safety, frontline supervisors' performance evaluations must include their ability to motivate employees to work safely. The instructor should detail the company's policy, which does not tolerate failure to use and maintain all PPE, especially respirators. Teaching new people to wear respirators comfortably, particularly in hazardous atmospheres, is similar to a test pilot constantly pushing the edges of the aircraft envelope. In this case, the instructor slowly and methodically pushes the respirator user to the edge of their comfort level with the unit. If the instructor is patient and the user is willing, the user's comfort level can continue to expand until the limitations of the respirator have been reached. An instructor who pushes the student too far or too quickly runs the risk of scaring the person, who may refuse to wear the respirator again. In that case, the time and effort involved to that point will have been wasted. The best approach is a training regimen that starts with respirator familiarity and progresses to more difficult and strenuous tasks as the student's confidence rises. Do not expect the entire response team or brigade to be at the same point in their abilities or confidence level. Physical fitness is imperative for maximum performance in emergency situations. A program that helps the employees maintain physical fitness is worth the effort.

Elaborate facilities and equipment for training is not always essential. Covering facepieces with wax paper and rubber bands enables the students to learn to communicate without seeing. Tables and chairs can be turned over to build a makeshift maze. Covering the student facepieces also allows the instructor to observe and correct mistakes quickly before they become bad habits. Performing searches in large industrial and commercial facilities is much more difficult than routine searches of small rooms and residential units. Pre-planning exactly how your response team or your local fire department would search your facility is part of good emergency planning. Infrared imaging devices may be necessary and special techniques may have to be developed to find trapped workers or reach the seat of the fire or spill. Remember the five Ps:

Pre-
Planning
Prevents
Poor
Performance

Finally, do not try to reinvent the wheel and train your workers by yourself. The industrial hygiene staff, the local fire department instructors, and fellow safety professionals may already have plans and programs that can be adapted to suit the facility. NIOSH, OSHA, the National Fire Protection Association, and other consensus standard bodies offer many excellent publications that can help with respirator training.

ELEMENTS OF A RESPIRATORY PROTECTION PROGRAM

The following excerpts from OSHA's general industry standard on respiratory protection outlines the basic elements of a written respiratory protection program.

Readers should know that this general industry standard is incorporated by reference in the construction and maritime regulations. Readers should reference the definitions, respirator selection, medical evaluation, fit testing, respirator usage, and other provisions in the standard for more specific requirements pertaining to the usage of respirators as a control for respiratory hazards.

1910.134(c)

Respiratory protection program. This paragraph requires the employer to develop and implement a written respiratory protection program with required worksite-specific procedures and elements for required respirator use. The program must be administered by a suitably trained program administrator. In addition, certain program elements may be required for voluntary use to prevent potential hazards associated with the use of the respirator. The Small Entity Compliance Guide contains criteria for the selection of a program administrator and a sample program that meets the requirements of this paragraph. Copies of the Small Entity Compliance Guide will be available on or about April 8, 1998, from the Occupational Safety and Health Administration's Office of Publications, Room N 3101, 200 Constitution Avenue, NW, Washington, DC, 20210 (202-219-4667).

1910.134(c)(1)

In any workplace where respirators are necessary to protect the health of the employee or whenever respirators are required by the employer, the employer shall establish and implement a written respiratory protection program with worksite-specific procedures. The program shall be updated as necessary to reflect those changes in workplace conditions that affect respirator use. The employer shall include in the program the following provisions of this section, as applicable.

1910.134(c)(1)(i)

Procedures for selecting respirators for use in the workplace.

1910.134(c)(1)(ii)

Medical evaluations of employees required to use respirators.

1910.134(c)(1)(iii)

Fit testing procedures for tight-fitting respirators.

1910.134(c)(1)(iv)

Procedures for proper use of respirators in routine and reasonably foreseeable emergency situations.

1910.134(c)(1)(v)

Procedures and schedules for cleaning, disinfecting, storing, inspecting, repairing, discarding, and otherwise maintaining respirators.

1910.134(c)(1)(vi)

Procedures to ensure adequate air quality, quantity, and flow of breathing air for atmosphere-supplying respirators.

1910.134(c)(1)(vii)

Training of employees in the respiratory hazards to which they are potentially exposed during routine and emergency situations.

1910.134(c)(1)(viii)

Training of employees in the proper use of respirators, including putting on and removing them, any limitations on their use, and their maintenance.

1910.134(c)(1)(ix)

Procedures for regularly evaluating the effectiveness of the program.

1910.134(c)(2)

The following indicate where respirator use is not required.

1910.134(c)(2)(i)

An employer may provide respirators at the request of employees or permit employees to use their own respirators, if the employer determines that such respirator use will not in itself create a hazard. If the employer determines that any voluntary respirator use is permissible, the employer shall provide the respirator users with the information contained in Appendix D to this section ("Information for Employees Using Respirators When Not Required under the Standard").

1910.134(c)(2)(ii)

In addition, the employer must establish and implement those elements of a written respiratory protection program necessary to ensure that any employee using a respirator voluntarily is medically able to use that respirator, and that the respirator is cleaned, stored, and maintained so that its use does not present a health hazard to the user. Exception: Employers are not required to include in a written respiratory protection program those employees whose only use of respirators involves the voluntary use of filtering facepieces (dust masks).

1910.134(c)(3)

The employer shall designate a program administrator who is qualified by appropriate training or experience that is commensurate with the complexity of the program to administer or oversee the respiratory protection program and conduct the required evaluations of program effectiveness.

1910.134(c)(4)

The employer shall provide respirators, training, and medical evaluations at no cost to the employee.

Review Questions

Note: Some of the following review questions require the reader to reference OSHA's standard on respiratory protection, 29 CFR § 1910.134, to obtain comprehensive answers.

1. In five sentences or less, distinguish the terms *gas* and *vapor.*
2. Why is it important for employees to have medical clearances before being permitted to wear certain types of respirators? Which respirators or under what circumstances are medical clearances required?
3. List the key OSHA-required elements of a respiratory protection program.
4. As a respiratory hazard, how do particulates affect the body differently than vapors, gases, and fumes?
5. What are critical atmospheric considerations when selecting a respirator?

4 Fire and Explosion Hazards

Even though the statistics reveal a remarkable reduction in the total number of fires, fire deaths, and fire loss costs since the first edition, America's fire death rates are too high for such a socially advanced and industrialized nation. The following, most-recent statistics from the U.S. Fire Administration* tell both an encouraging and yet troubling message about fires in the United States.

- In 2014, the total number of fires in the United States was down 22.7 percent. Still, there were 1,298,000 fires in that year.
- The number of deaths from fire in the United States was down 11.7 percent in 2014 when compared with data from 2005. Despite this reduction, there were still 3,275 fire deaths in 2014.
- The amount of fire loss in 2014 was a 21-percent reduction from the loss experienced in 2005. Still, fire loss cost Americans $11.6 billion in 2014 alone. The following statistics published by the National Fire Data Center illustrate the size and scope of the problem.

The United States has a severe fire problem, more so than is generally perceived. Nationally, there are millions of fires, thousands of deaths, tens of thousands of injuries, and billions of dollars lost—which makes the U.S. fire problem one of great national importance.

THE FIRE PROBLEM IN NONRESIDENTIAL/COMMERCIAL SETTINGS

Much of the effort in fire prevention, both public and private, has gone into protecting nonresidential structures, and the results have been highly effective—especially when compared with the residential fire problem. Between 1994 and 1998, nonresidential structures represented approximately 4 percent of fire deaths, 10 percent of fire injuries, 31 percent of total fire dollar loss, and 8 percent of all fires. Collectively, heating systems/efforts, electrical malfunctions, exposure to open flames and heat sources, equipment malfunctions, and careless behavior account for just under 50 percent of non-residential fires. Many of these causes are directly related to fire hazards for which employers have either enacted controls or should enact controls (hot work permitting, preventive maintenance, training, etc.).

---

* See https://www.usfa.fema.gov/data/statistics/#tab-1.

DEALING WITH THE FIRE PROBLEM

To deal effectively with the fire problem, a strategy must be developed. The following steps should be taken to minimize the negative impacts of hostile fires:

1. Assess the potential for fire.
2. Maximize fire prevention efforts.
3. Utilize early detection methods.
4. Utilize automatic suppression systems.
5. Slow the growth of potential fires.
6. Prepare a fire preplan and train employees in it to minimize risk and maximize safety and property protection.

STEP ONE: ASSESS YOUR POTENTIAL FOR FIRES AND EXPLOSIONS

Take a thorough walk through the facility to determine how much combustible material you have and the proximity of those combustibles to ignition sources. At the same time, try to determine what natural or artificial boundaries exist due to building construction and compartmentalization. Are materials with tremendous heat release and ease of ignition isolated from potential ignition sources? If building compartments exist, are they fire rated with suitable fire-rated closures to protect all openings?

Does your facility have any unique fire protection problems that require special extinguishing systems or special extinguishing agents? If these systems exist, have they been inspected in accordance with the appropriate section of the National Fire Protection Association (NFPA) Code? If your facility is protected by an automatic sprinkler system, was it designed, installed, inspected, and maintained based upon the current fire load and occupancy use? Are there any areas of the facility that are not protected by automatic sprinklers? Are the private and public water supply systems adequate and maintained to ensure reliability and dependability? Have the public and private hydrants been flowed and maintained regularly? Are all control valves on all fire protection equipment of the indicating type? Are all the control valves electronically supervised or chained and locked in the open position?

Contact your insurance carrier to ask for the insurance service office (ISO) rating of all the fire departments near your facility. The ISO rating schedule and system are used to provide information for insurance companies regarding a particular fire department's ability to extinguish fires. Class 1 indicates the highest level of protection, while Class 10 reflects little or no capability. Use this information as a starting point to evaluate the ability of local response agencies to help you in the event of a fire. What is their response time to your facility? Most ISO Class 1 fire departments will have sufficient staffing to consistently arrive at your facility within a prescribed time based upon their distance from your facility. Many communities in America are protected by volunteer fire departments whose response time and staffing level may vary depending upon the time of day. Invite the local fire department in for a tour of your facility. Ask them questions to determine how knowledgeable they are about the facility and, more important, how eager they are to learn and accept the challenges of protecting your facility. The bottom line here is that if you have a solid, Class 1 fire department located across the street, you may only have to provide them with technical support and technical assistance in the event of a fire. However, if you have a Class 9 or 10 fire department with a 15-minute or longer response time, you will have to rely on the facility's automatic fire protection and whatever level of manual suppression force (fire brigade) you can assemble, train, and equip. All too often company risk managers will tell new safety managers, "You never want to start a fire brigade because they are too expensive." The decision to establish a fire brigade cannot be made in a vacuum. If you are faced with long response times or questionably trained and equipped response agencies, a fire brigade may be considerably cheaper than having to rebuild the facility and attempt to regain the customer base lost while your facility was out of business.

STEP TWO: MAXIMIZE FIRE PREVENTION EFFORTS

To the extent practical and possible, try to keep the items that can burn away from the sources that can ignite them. This is especially true when dealing with materials that are easily ignited or have an unusually high fire load or rapid fire growth. The very nature of many industrial operations prohibits complete isolation of combustibles and ignition sources so your only choice is to minimize the exposure by building storage rooms for dangerous commodities and keeping no more than a one-day supply in the area where the ignition sources are present. Consider facility design features such as maximum foreseeable loss (MFL) walls, which limit fire spread, and ensure that the automatic sprinkler protection is adequate for the hazard. Make sure that sprinklers are inspected and maintained according to the appropriate sections of the NFPA fire code. Advice concerning standard sprinklers is found in NFPA 13, but storage occupancy information, including sprinkler design, can be found in the appropriate NFPA 230 series fire code based upon the material being stored. Educate your people and help them understand the implications of a major fire; for example, some facilities that suffer a major fire may not rebuild in that same location, as the new facility may be moved to a part of the world where labor or energy costs are lower. Hold supervisors accountable for the fire prevention efforts of their employees in their performance evaluations.

STEP THREE: UTILIZE EARLY DETECTION METHODS

Make sure that your facility is protected throughout with an appropriate fire alarm system. Use the early-warning type of detectors where practical. If explosion hazards exist, the system will have to be fast enough to sense the explosion and react to discharge the extinguishing agent after the explosion starts but before the pressures rise to damaging levels. These exotic systems are custom-designed and installed for each hazard they protect against. They are routinely used in conjunction with explosion venting. Advances in laser detectors, multiple sensing detectors, and air sampling systems are designed to provide early detection and warning while minimizing false alarms. Proper inspection, testing, and maintenance will ensure that the system is operational and will minimize false non-fire activations. Consult the NFPA

72 series for information regarding fire alarm system design, installation, inspection, and maintenance.

Step Four: Utilize Automatic Suppression Systems

Since the 1800s and the advent of the automatic sprinkler head, statistics have proven that sprinklered buildings are much less likely to be destroyed by fire. Insurance companies recognize this and give substantial discounts to facilities that choose to install, inspect, and maintain automatic fire sprinklers. If the occupancy use or fire load have increased since the original sprinkler system was installed, new sprinkler heads with larger diameter orifices may be able to increase the density (gallons per minute per square foot). If a hydraulic analysis of the system proves the existing sprinkler piping and water supply will accommodate the larger diameter heads, the system can be upgraded by switching the heads. In other cases, it may be necessary to replace some or all of the piping to upgrade the fire protection system. While NFPA 13 provides guidance for standard sprinkler systems, storage occupancies have additional requirements listed in the individual storage standards of the NFPA 230 series. Many industrial installations have sufficient quantities of flammable liquids present to merit the installation of automatic or manually activated foam–water systems for controlling Class B fires. These must be maintained in accordance with the appropriate NFPA 11 or 11A. If water in sufficient flow and pressure is not available, one or more fire pumps may be necessary. These pumps must be installed, inspected, tested, and maintained under the regulations of NFPA 20.

Step Five: Slow the Growth of Potential Fires

Recognizing that fire prevention efforts will never be 100 percent successful, it is necessary to make plans in the event that a fire does occur. The National Fire Protection Association lists the following approaches or strategies to limit the growth of fires after they start*:

1. Restrict materials used in contents and furnishings to reduce heat release, reduce the smoke-generation rate, and prevent unusually high amounts of toxic materials relative to the quantity of smoke generated.
2. Add fire retardant to materials to slow growth and heat release.
3. Use fire-resistive barriers to slow the spread of fire to large secondary items.
4. Restrict fuel load by limiting contents based on total fuel potential.
5. Restrict linings, wall coverings, ceiling coverings, and floor coverings of rooms to prevent rapid flame spread.
6. Restrict the use of combustible materials in concealed spaces.
7. Require safe handling of large quantities of potential fuel.

---

* *Fire Protection Handbook*, 17th ed., National Fire Protection Association, Quincy, MA, 1991.

STEP SIX: PREPARE A FIRE PREPLAN AND TRAIN EMPLOYEES TO MINIMIZE RISK AND TO MAXIMIZE LIFE SAFETY AND PROPERTY PROTECTION

The Occupational Safety and Health Act gives employers three choices regarding their response to fire emergencies.

1. First, an employer may develop an emergency action plan that requires all employees to evacuate when a fire emergency occurs. In this case, employees are not assuming any fire extinguishing tasks. While this option provides maximum protection to employees—provided all employees follow the plan—it is sometimes not practical or wise with respect to the business operations. The employer's emergency action plan must specify evacuation procedures and train employees on how to report fires and sound the evacuation alarm. The employer must also post evacuation routes and train employees on exiting procedures.
2. An employer may also elect to have designated employees perform specific firefighting duties as trained and equipped. These duties could range from fighting only small, incipient-level fires with extinguishers or small hose lines up to interior structural firefighting. Again, the training requirements and equipment must be consistent with the expected level of response in the fire brigade mission and operating procedures.
3. The employer may choose to permit all employees to engage in firefighting duties consistent with their training and equipment. This option is usually reserved for incipient-stage brigades, and the employees are responsible for using extinguishers or small hose lines to extinguish small fires in their work areas.

In the case of interior structural brigades, the training and equipment are equivalent to their municipal firefighting counterparts. In addition to training the brigade members, the level of training for the brigade officers increases with increased levels of response. If your brigade has employees who also belong to response agencies outside their employment with your company, their training with the outside agencies can count toward required brigade training. The experience these individuals bring to the brigade often enables them to lead, motivate, and train other brigade members.

Occasionally, someone will say, "I don't have to comply with fire brigade training and equipment requirements because our team is a Hazardous Materials Response Team" (or some other name). It does not matter what the team is called. If the mission and operating procedures allow them to use fire protection equipment for controlling fires, the team must be trained and equipped in a manner consistent with that level of response. Utilizing a combined response team makes good sense. Nearly all emergency response team members must have annual physicals to ensure their ability to perform strenuous work and wear respirators. They must understand how to function under an incident command structure and have at least hazardous materials awareness level training. Many companies combine fire, Hazmat response, emergency medical care, and confined space entry/rescue under a single team with individuals qualified at various levels depending upon their training and abilities. Another often overlooked advantage of spending the time and money to train and equip a response team is the ability of team members to recognize hazardous situations and mitigate them before a disaster occurs.

Just as important as the options for fire response, OSHA sets forth a requirement for employers to have a fire prevention plan. The requirements for a general industry fire prevention plan are set forth in 29 CFR § 1910.39, and are as follows.

1910.39(a)

Application. An employer must have a fire prevention plan when an OSHA standard in this part requires one. The requirements in this section apply to each such fire prevention plan.

1910.39(b)

Written and oral fire prevention plans. A fire prevention plan must be in writing, be kept in the workplace, and be made available to employees for review. However, an employer with 10 or fewer employees may communicate the plan orally to employees.

1910.39(c)

Minimum elements of a fire prevention plan. A fire prevention plan must include the following.

1910.39(c)(1)

A list of all major fire hazards, proper handling and storage procedures for hazardous materials, potential ignition sources and their control, and the type of fire protection equipment necessary to control each major hazard.

1910.39(c)(2)

Procedures to control accumulations of flammable and combustible waste materials.

1910.39(c)(3)

Procedures for regular maintenance of safeguards installed on heat-producing equipment to prevent the accidental ignition of combustible materials.

1910.39(c)(4)

The name or job title of employees responsible for maintaining equipment to prevent or control sources of ignition or fires.

1910.39(c)(5)

The name or job title of employees responsible for the control of fuel source hazards.

1910.39(d)

Employee information. An employer must inform employees upon initial assignment to a job of the fire hazards to which they are exposed. An employer must also review with each employee those parts of the fire prevention plan necessary for self-protection.

SUMMARY

While tremendous strides have been made with both fire prevention and fire response over the last several decades, there remains a remarkable problem with fires in the workplace. Ultimately, the employer's fire prevention plan and efforts will keep a fire from occurring. Nonetheless, the employer must evaluate its capabilities and preplan its response to a fire. In preplanning a fire response, the employer should evaluate the extent to which it will engage—if it engages at all—in fire suppression. Regardless of whether an employee's role is simply to evacuate a facility or perform fire suppression tasks as part of an industrial fire brigade, all must be trained on their expected role.

Review Questions

1. List at least eight hazard controls that would prevent fires in the workplace.
2. What are an employer's choices when deciding how to respond to fires?
3. Where, or to whom, may a safety professional look for guidance when developing fire prevention and response plans?
4. What are the major elements of a fire prevention plan as required by OSHA?
5. What are five things that can be done to mitigate the effects of a fire?

5 Confined Space Hazards

For many years, employees were being killed or injured while working in areas not designed for continuous human occupancy. Even worse, more than one half of the fatalities in these spaces were rescuers. In a typical scenario, an employee entered the space to inspect, repair, or perform routine maintenance in the space. Either because the conditions in the space were not checked before entry or because conditions deteriorated after the employee entered the space, the original employee failed to emerge from the space. Coworkers or even emergency response personnel all too often enter the space in an attempt to rescue the worker and become fatalities themselves. Although asphyxiation is one of the primary causes of death, other causes include engulfment, entanglement with moving mechanical equipment, electrocution, fires, explosions, and toxic gases and corrosive liquids. OSHA's response to the problem has been a proactive approach. 29 C.F.R. 1910.146 is concerned with the dangers of the confined spaces, but this regulation mandates the use of an administrative control (permit) system to eliminate all of the potential hazards a worker could encounter before the worker enters the space. If the permit-required confined space (PRCS) program is operating effectively, it should prevent entry into an unsafe space and require the entrants to leave the space if conditions deteriorate while they are still in that space. The standard also provides additional requirements for equipping, training, and evaluating rescue teams if an emergency should arise in a permit-required confined space. Even if an employer chooses to prohibit employees from entering a PRCS, the employer still must educate employees about the dangers of that PRCS. The employer must also post signs at each PRCS warning employees of the danger and prohibiting them from entering the PRCS. One of the first questions a new safety manager asks when learning of the PRCS standard is, "What is a confined space and are there any in my facility?"

DEFINING CONFINED SPACES AND PERMIT-REQUIRED CONFINED SPACES (PRCS)

According to OSHA, a *confined space*:

1. Is large enough and so configured that an employee can bodily enter and perform assigned work.
2. Has limited or restricted means for entry or exit (for example, tanks, vessels, silos, storage bins, hoppers, vaults, and pits are spaces that may have limited means of entry).
3. Is not designed for continuous employee occupancy.

A *permit-required confined space* means a confined space that has one or more of the following characteristics:

1. Contains, or has the potential to contain, a hazardous atmosphere.
2. Contains a material that has the potential for engulfing an entrant.
3. Has an internal configuration such that an entrant could be trapped or asphyxiated by inwardly converging walls or by a floor which slopes downward and tapers to a smaller cross-section.
4. Contains any other recognized serious safety or health hazard.

Although many confined spaces are enclosed, there are also open-top pits and tanks that meet the definition of both a confined space and a PRCS. Consultants are frequently asked to provide a list of all the equipment necessary for confined space entry and rescue teams; however, it is not possible to determine what equipment will be required until each space has been studied in detail to assess all of the hazards and the best countermeasures to prevent injury.

CLASSIFICATION OF HAZARDS

RESPIRATORY HAZARDS

Because asphyxiation is one of the leading causes of death and injury in confined spaces, everyone working in or near a PRCS must understand these hazards. OSHA does not permit anyone to enter or occupy a space containing less than 19.5 percent oxygen. Even though humans may be able to hold their breath for a couple of minutes, a single breath of an atmosphere without oxygen will result in immediate unconsciousness and death if the victim is not removed to fresh air within minutes. OSHA standard 29 C.F.R. 1910.146 requires monitoring of the space to ensure that at least 19.5 percent oxygen is available prior to and during any confined space entry. Simple asphyxiants such as nitrogen, argon, carbon dioxide, and any other gas or vapor that can accumulate and displace the oxygen can be fatal in a PRCS.

Toxic respiratory hazards could be present even if sufficient oxygen is available. Toxic atmospheres involve substances that even in very low concentrations can kill or injure workers. Often, industrial processes utilize or have toxic by-products. Some common examples include hydrogen sulfide, sulfur dioxide, chlorine, and carbon monoxide. Refer to the Material Safety Data Sheet (MSDS) or other sources to determine the permissible exposure limit (PEL) for the chemical identified. Specific test equipment must be used to analyze all the potential toxins that may be present in a PRCS. The PRCS must be purged until the level of toxins present is within acceptable limits. Personal protection equipment (PPE), including a self-contained breathing apparatus (SCBA), should be worn to minimize the risk of injury should the concentration of toxins rise before the entrants can escape. If the hazard cannot be identified or the concentration of the hazard cannot be quantified, a positive-pressure, self-contained breathing apparatus is the only acceptable choice. If the hazard and concentration can be determined, an appropriate respirator and PPE should be selected to ensure that the entrants never exceed the PEL for that agent.

CORROSIVE HAZARDS

Corrosive atmospheres have the ability to cause damage to the skin or any other human tissue they contact. Strong acids and alkali materials are used in many industrial processes and are by-products generated in others. These atmospheres also have the ability to damage or destroy personal protective equipment. All PPE selected for use in corrosive atmospheres should be compatible with the materials in the space. Some SCBA manufacturers offer special elastomers for use in chlorine environments.

ENERGY HAZARDS

Because confined spaces by definition are not designed for continuous human occupancy, they often lack the physical space, adequate lighting, and other creature comforts that humans demand. These same hazards often conceal other deadly hazards such as energized electrical components, high-pressure gas and liquid transmission lines, and pneumatic or hydraulic equipment and components. The only way to protect entrants from these hazards is to educate them about all the hazards in the space and by isolating and de-energizing every component in the space before allowing them to enter.

ENGULFMENT (CONTENTS) HAZARDS

Another type of PRCS hazard that has claimed lives is engulfment. Failure to isolate valves that direct contents into vessels is one of the problems. Unauthorized entry into open-top containers with sloping sides and containing grain or other similar materials has also resulted in fatalities. One of the primary ways to guard against unauthorized entry and improperly compensated risk is through effective education. Every employee must be taught to appreciate the dangers, and they must understand the company policies that will not tolerate unauthorized entry.

FIRE AND EXPLOSIVE ATMOSPHERIC HAZARDS

Flammable atmospheres pose serious fire and explosion threats to PRCS entrants. The permit system requires the space to be monitored before entry and continually to ensure that flammable vapors do not accumulate while the entrants are inside the space. If a reading of 10 percent or greater of the lower flammable limit of the gas is reached according to the combustible-gas monitoring instrument, the entrants must exit the space. The probe of the monitoring instrument should be located in the space where the vapor is most likely to be located. For example, most petroleum product vapors are heavier than air and will accumulate near the bottom of enclosures.

MECHANICAL EQUIPMENT HAZARDS

Mechanical devices in many PRCSs are designed to mix, chop, blend, and stir products. If the primary energy to drive the devices is not properly locked out, according to

29 C.F.R. 1910.147, and if the energy stored in devices such as accumulators, capacitors, compressed air tanks, and so on is not bled off, equipment may activate and injure or kill the entrants. It is worth noting that a single machine may have several different energy sources. For example, a woodshop surfacing machine may have a 220-volt power circuit for the cutting head motor, a 110-volt circuit to control the motors that position the feed table, and a 12- or 24-volt DC control circuit to operate the machine controls. All three sources must be checked to verify that no energy is present.

FALL HAZARDS

The Occupational Safety and Health Administration requires all employees to be protected from falling from any surface four feet (29 C.F.R. 1910.23) to six feet (29 C.F.R. 1926.501) above adjacent surfaces. A PRCS is not exempt from this duty to prevent falls. Each employee on a walking or working surface (horizontal or vertical surface) with an unprotected side or edge that is six feet (1.8 meters) or more above a lower level must be protected from falling by the use of guardrail systems, safety net systems, or personal fall-arrest systems. If the only access to a PRCS is by lowering the employee on a rope, a personal fall-arrest system may be the only practical choice.

ENVIRONMENTAL CONDITION HAZARDS

Other factors such as extreme temperature may compound the hazards present in the PRCS. Steam pipes or HVAC pipes may be hot or cold enough to cause burns to the skin, or the temperature in the space may be hot enough to cause fatigue and collapse or cold enough that frostbite of extremities may be a concern. Any condition that has the potential to distract the entrant from focusing on the other safety hazards could contribute to an accident or injury.

Because many PRCSs are dark, tiny spaces, encountering reptiles, rodents, or insects is another potential hazard. In such a situation thermal imagers could possibly be used, as any living animal will give off heat and be visible even in low- or no-light conditions. Noise and vibration are environmental conditions that complicate work or rescue in a PRCS. Whenever possible, machinery near the PRCS should be stopped to reduce excess noise and vibration and facilitate communication.

Psychological hazards of working in a restricted space can adversely affect workers. Individuals who are susceptible to claustrophobia could have problems before or during entry into confined spaces. Employers should attempt to uncover this tendency through training prior to working in actual hazardous conditions. People with minor or moderate claustrophobia anxiety in these working conditions can often be slowly acclimated through training and can learn to overcome the anxiety. Individuals who cannot do so should not be forced to work in a PRCS, because they could create dangerous conditions for all entrants.

COMMUNICATION ISSUES

Communication problems compound the already difficult work conditions in many PRCSs. SCBA manufacturers are beginning to integrate communication systems into the confined space SCBA equipment. These hardwire systems bundle communication wires with the supplied air hose of systems that enable the workers or rescuers to wear smaller escape SCBA cylinders while still remaining in the space long enough to accomplish the assigned task (work or rescue). The voice-activated microphones and speakers in these systems allow workers to communicate with each other and with support personnel outside the space. In high-noise environments, it may be necessary to wear headsets to facilitate communication.

In the absence of high-technology communication systems, other methods of communicating can be used with varying degrees of success. The use of a rope with a system of coded tugs has proved effective in some situations. If the space is small enough and quiet enough, a worker outside may be able to talk to the worker in the space without assistance. Workers in tanks have developed systems of tapping on the tank to communicate with the PRCS attendant on the outside. The OSHA standard requires the attendant to remain in communication with the entrant.

Entry and exit from a PRCS often presents significant challenges to the entrants and to rescuers trying to remove victims. Whenever possible, entrance to the PRCS should be by means level with the ground or within a few feet above the ground. All too often this is not possible. Each PRCS space must be evaluated to design adequate means for entering and exiting the space. Many companies manufacture devices designed to facilitate lowering and raising entrants and rescuers into a PRCS. Any equipment purchased for PRCS rescue must be rated for lifting humans. Workers who are lowered into spaces must keep retrieval lines on themselves while in the space unless doing so increases the danger.

Physical conditions inside the space may demand that the biggest, strongest employee performs the work or rescues the worker, but the arrangement of the space may only permit entry of the smallest member of the team. Pipes, wires, hoses, and equipment can entangle an entrant, causing frustration at best and panic at worst. The presence of grease, oil, water, steam, or other products only adds to the difficulty. Although already mentioned under mechanical hazards, entrants may find live electrical wires in many PRCSs. Do not overlook these hazards when evaluating any PRCS. Training for rescue in spaces with the physical hazards mentioned in this section would benefit from the availability of props designed to simulate conditions in the PRCS. Workers can learn without the actual dangers present and the instructor can observe and provide input and assistance.

ATMOSPHERIC TESTING AND MONITORING

When selecting equipment to test and monitor confined spaces, several factors should be considered. Colorimetric style detector tubes may be fine for isolating and determining the identity and concentration of unknown substances, but they are only a snapshot analysis of the space at the time the sample was drawn into the tube. The conditions in the space may have already changed by the time the color change has been interpreted. Equipment specific to the hazardous substances identified should be utilized in a manner that continuously monitors the conditions in the space for the presence of, for example, oxygen, flammable/explosive vapors, or any possible toxic gases. These readings should be recorded on the PRCS permit before entry and periodically during the operation. Audible and visual warnings should announce hazardous conditions even if the attendant monitoring the space is distracted by some other activity. The equipment selected must be simple, reliable, and easily understood and operated by the PRCS entry and rescue team members. Individual sensors in monitoring equipment can fail at any time, so it is important to rely on suppliers who can service your equipment quickly and to have additional spare sensors and monitors available. When a piece of equipment monitoring the PRCS fails, everyone in the space must exit and remain out of the space until functioning equipment is back on the job ensuring their safety.

ENTERING A PRCS

At a minimum, the following positions are required by the PRCS standard.

ENTRY SUPERVISOR

This supervisor is responsible for ensuring that all aspects of the permit have been addressed and that all hazards have been controlled and documented on the appropriate portions of the entry permit. The permit shall be signed by the entry supervisor, verifying the accuracy of information on the permit.

ENTRANTS

These individuals will have witnessed the posting of, or had the opportunity to see, the permit indicating all the hazards identified and the measures taken to control them. The entrants must know the hazards in the space and be able to recognize signs, symptoms, and consequences of exposure to the hazards. They will wear all PPE required to work safely in the space and communicate with the attendant outside the space. They are also responsible for notifying the attendant of any changes in the space and for exiting as quickly as possible if a prohibited condition is detected or if the attendant orders an evacuation of the space. Whenever possible, the entrants should work in teams of two; however, some PRCSs will not have sufficient room for more than one person to enter.

ATTENDANT

This individual must maintain communication with the entrants. The attendant may monitor more than one PRCS if doing so does not diminish the ability to monitor and protect all the entrants in all of the PRCSs that the individual is responsible for monitoring. The attendant duties include the following:

- Monitoring conditions inside and outside the space to determine if it is safe for entrants to remain in the space
- Warning unauthorized persons to stay away from the PRCS
- Performing non-entry rescues as specified in the employer's rescue procedure
- Summoning rescue and emergency services as soon as the attendant determines that the entrants may need assistance to escape the PRCS

Attendants may only enter as members of confined space rescue teams if they are properly trained and equipped and after they have been relieved of their attendant duties. Attendants may not engage in any activities that reduce their ability to monitor and protect the entrants in the PRCSs that they are responsible for monitoring.

RESCUERS

Statistics reveal that more than one half the fatalities in confined space incidents are individuals involved in attempting to perform rescues. Procedures must be developed and enforced to prohibit rescue by anyone other than those people trained and equipped. In some cases, PRCS rescue may be delegated to outside or off-premise personnel. If the victims are to have any chance for survival, the rescue team response times must be quick (minutes). If outside-agency personnel are used for PRCS rescues, the agency selected should be advised of specific hazards in those spaces. Additionally, OSHA has updated its PRCS standard and now sets forth the following requirements when employers designate rescue and emergency services:

- Evaluate a prospective rescuer's ability to respond to a rescue summons in a timely manner, considering the hazard(s) identified. (*Note*: What is considered "timely" will vary according to the specific hazards involved in each entry. For example, §1910.134, Respiratory Protection, requires that employers provide a standby person or persons capable of immediate action to rescue employee(s) wearing respiratory protection while in work areas defined as IDLH atmospheres.)
- Evaluate a prospective rescue service's ability, in terms of proficiency with rescue-related tasks and equipment, to function appropriately while rescuing entrants from the particular permit space or types of permit spaces identified.
- Select a rescue team or service from those evaluated that
  - Has the capability to reach the victim(s) within a time frame that is appropriate for the permit space hazard(s) identified.
  - Is equipped for and proficient in performing the needed rescue services.
- Inform each rescue team or service of the hazards they may confront when called on to perform rescue at the site.
- Provide the rescue team or service selected with access to all permit spaces from which rescue may be necessary so that the rescue service can develop appropriate rescue plans and practice rescue operations. (Note to paragraph (k)(1): Nonmandatory Appendix F contains examples of criteria which employers can use in evaluating prospective rescuers as required by paragraph (k)(l) of this section.)

PERMIT-REQUIRED CONFINED SPACE RESCUE PRIORITIES

The incident commander of the PRCS rescue team must quickly assess the scene and categorize the rescue as one of three different priorities.

Priority One

Priority one involves rescue from life-threatening situations from which the entrants must be removed quickly or they will die. The cause of the emergency could be an injury to a worker in the space or a change in the environment, or both. In this scenario, all medical treatment must be delayed until the entrants are removed from the hazard area. This is termed *snatch and drag* by some rescue technicians. It is hoped that the entrants still have their retrieval lines attached and that the attendant has removed one or more of the entrants prior to arrival of the rescue team.

Priority Two

In this scenario, the victim is disabled by a non–life-threatening injury or illness while inside the PRCS. Atmospheric monitoring indicates that environmental conditions inside the space are still safe but the victim requires medical assistance or transport from the space. Because the urgency is reduced, the rescuers take the time required to treat the patient, then package the patient for transport from the PRCS and finally remove the patient from the space. This may require special equipment and techniques to immobilize the patient and protect them from further injury. The rescue team members must have the medical expertise to effectively perform first aid as required. There may be situations where the illness or injury may dictate some level of urgency, but the environmental conditions in the space do not present a risk to the rescuers.

Priority Three

In this scenario, victims are found in a PRCS and the initial rescue size-up (assessment of the scene) reveals that anyone remaining in the space at this time could not have survived. This is a difficult determination to make; however, failure to do so places the rescue team members in a "life-swapping" situation. If the victims in the space have no chance of survival, it makes no sense to risk further lives for body-recovery operations. Remember that over one half of the fatalities in PRCS rescues are would-be rescuers. If there is a potential life to be saved, then the situation is not priority three; it is priority one.

Permit-Required Confined Space Training

Permit-required confined space training falls into three categories. After successful completion of PRCS awareness training by all newly hired employees, the employee or independent contractor will be able to

1. Recognize PRCSs by the posted signs and by OSHA/company definitions.
2. List the location of the PRCS nearest their work area(s).
3. Be able to accurately describe the PRCS policies concerning permits, approvals, entry, and rescue.

Confined space entry training is provided to everyone who may serve in any capacity in a confined space entry. The training includes all elements of the PRCS entry permit system and approval process, types of hazards and dangers, countermeasures, atmospheric monitoring, authorization, and entry team member duties and responsibilities. The training also contains site-specific hazard and permit information for each PRCS identified in the facility. The nature of each PRCS presents different hazards and unique equipment requirements to enter and work safely. For example, some PRCSs require horizontal entry while others require lifting harnesses, ropes, and cranes to facilitate entry and exit.

PRCS rescue team member training starts with confined space entry training and builds upon it. In addition to a working knowledge of safe ways to work in all of the PRCSs of the facility, the rescue team members are given additional training as follows:

- They are made aware of common PRCS safety violations and the historical consequence of those violations.
- They receive training in emergency medical care consistent with the level of response and care the team will be expected to provide; first aid and CPR are considered the minimum training for basic life support.
- They receive training on inspection, operation, and maintenance of every piece of PRCS entry and rescue equipment they will have available. This would include ropes, lifting devices, tripods, harnesses, immobilization equipment, self-contained breathing apparatus, atmospheric monitoring equipment, and communication and incident command procedures.
- In some companies, the emergency response organization may utilize some or all of the same people on both the fire brigade and PRCS team. This facilitates suppression of fires to effect the rescues more quickly, but the commander must ensure that, prior to rescue team entry, none of the rescue team members is involved in firefighting operations. PRCS rescue team members must focus on their roles to protect the rescue team. Backup rescue team members must be ready to enter immediately if the primary rescue team has problems and must be removed.
- The rescue team may enter immediately dangerous to life and health (IDLH) atmospheres to perform rescues, so it is imperative that competency-based testing be conducted to demonstrate that every team member has the knowledge, skills, and attitudes necessary to work safely and not endanger other team members. Skill sheets should allow checking-off of all critical steps associated with the various pieces of equipment and procedures.
- Site-specific training on each PRCS will include identification of all hazards associated with each PRCS and the safest way to effect each type of rescue from that particular confined space. The initial training should focus on preparation for the most frequently entered spaces and the PRCSs that have already been involved in dangerous incidents or have the greatest potential for injuries or fatalities. Practice is the key to building team skills and confidence.

- Sometimes training will reveal rescue team members who suffer from claustrophobia. Often this can be dealt with effectively by slowly introducing the person to a controlled, safe, monitored confined space a little at a time and then allowing the person to stay in the space a few more minutes with each subsequent training session. This method has enabled individuals with extreme anxiety to overcome their fears. In the event that a team member cannot overcome these fears, they must not be permitted to serve as primary rescue team entrants or even as backup rescue team members. They can still serve in many roles outside the space, such as incident commander, atmospheric monitor, rigger (person who erects the necessary lifting devices), or hoist man to raise victims or pull them from the PRCS.
- The incident command system is the modular command and control system used to manage all emergency response incidents. Teaching the PRCS team to understand and work under this system brings direction, order, and control to the scene and allows the rescue commander to maintain an effective span of control in the event that additional resources are required. This also enables the team to work effectively with outside-agency personnel if they are called to assist.

ELEMENTS OF A WRITTEN CONFINED SPACE PROGRAM

1. The program must be in writing.
2. It must identify all confined spaces in the facility.
3. The program must develop and implement a written entry permit system.
4. The program must address the techniques for monitoring the confined space atmosphere before and during the entry.
5. Personnel must be selected, trained, and equipped with the necessary equipment for confined space entry and rescue.
6. Protective equipment must be acquired and entry and rescue team members must train with it prior to entry and/or rescue operations.
7. Procedures must be developed and training provided to ensure that attendants and emergency response team members are competent.
8. This program should be reviewed annually and updated every time a PRCS changes or new ones are added. All elements should be included in the evaluation, including training, equipment, the permit system, and hazard identification of all PRCS.

ELEMENTS OF A PRCS ENTRY PERMIT

1. Date, time, and location of the PRCS to be entered.
2. Section for recording the signatures of entrants, attendants, and entry supervisor.
3. List of all the hazards in the PRCS being permitted.
4. Corresponding list of all the countermeasures taken to eliminate each of the hazards identified with this PRCS.

5. Means for verifying that pre-entry conditions are in compliance with minimum acceptable entry conditions.
6. Place on the permit for recording initial and periodic monitoring
7. Equipment available at time of entry.

Review Questions

1. Distinguish a permit-required confined space and a confined space.
2. What are some notable atmospheric hazards that may be encountered in a confined space?
3. If an employer elects to designate an outside source to perform confined space rescuer duties, explain, in basic summary terms, what the employer must do.
4. What are the key elements of a PRCS entry program?
5. If you had to choose one PRCS entry role as most crucial, what role would it be and why?

6 Electrical Hazards

One of the most significant hazards presenting significant risk of fatal or catastrophic consequences is exposure to electricity. To address the various electrical hazards in the workplace, the Occupational Safety and Health Administration (OSHA) has developed several standards primarily governing the use of electricity in the industrial environment. In the area of installation and maintenance, the National Electric Code (NEC) is the primary set of regulations. Safety professionals should be aware that the National Electric Code has been adopted, in part, within the OSHA standards and is often adopted by various state and local agencies, especially in the area of building codes and construction codes. To protect workers who must be exposed to live or energized electrical circuits and equipment to perform their work from arc flash/ blast incidents and shock/electrocution, OSHA's stand on electrical safety–related work practices in conjunction with the National Fire Protection Administration's standard on the same hazard (NFPA 70E) should be followed closely.

This chapter addresses several potential electrical hazards, as well as general safeguards. Given the nature and importance of the OSHA Control of Hazardous Energy standard and its relationship to electrical energy, this standard will also be discussed.

ELECTRICAL SAFETY

In a recent study, it was determined that 1,000 people are electrocuted yearly in the United States, most from low-voltage current (see Table 6.1). This issue is a major concern for industry, which must provide the proper equipment and education to protect employees from these hazards.

GENERAL RULES FOR OPERATING AROUND ELECTRICITY

The following are recommendations for providing a safe environment while working around electricity.

EMPLOYERS

- Employers must promulgate and enforce a set of safety rules for their employees that is consistent with the conditions at their workplace.
- The employer will use positive pressure to ensure that employees observe the rules.
- Employers will conduct training sessions for employees so they will know proper, safe procedures.
- Access to hazard locations will be restricted to qualified employees.
- Employers must provide employees with easy access to safety equipment as needed.

TABLE 6.1

Effects of 60-Hz Current on an Average Human

Current Values	Effect
Safe Current Values	
1 milliampere or less	Causes no sensation; cannot be felt; at threshold of perception
1–8 milliamperes	Sensation of shock; not painful; muscular control temporarily lost
Unsafe Current Values	
8–15 milliamperes	Painful shock; can let go at will; muscular control not lost
15–20 milliamperes	Painful shock; cannot let go; muscular control of adjacent muscles lost
20–50 milliamperes	Painful; severe muscular contractions; breathing difficult
100–200 milliamperes	Ventricular fibrillation, a heart condition that results in death
200+ milliamperes	Severe burns; muscular contractions so severe that chest muscles clamp heart and stop it for duration of shock

EMPLOYEES

- Be alert.
- Be cautious.
- Know your job.
- Observe the rules.
- Inspect.
- Identify.
- Isolate.

At no time should employees perform jobs they have not been trained to do!

PLANNING

GENERAL

- All work performed on a facility's electrical system must be documented in writing.
- The electrical department will assist where needed to keep the plant's online diagram and panel directories up to date.

WORK PLAN

- Prework discussions should be held before each job by the responsible supervisor with the employees involved.
- When more than one crew is working at the same location or on the same circuit, the supervisors of all crews should meet and exchange information regarding each crew's duties.
- The senior supervisor on the multicrew job is responsible for the safety of the entire job.

WORK CREWS

- Work performed on live voltage should only be performed by a qualified, experienced employee.
- No employee should work alone on a circuit energized above 300 volts.
- No employee should work alone on a circuit energized above 30 volts when the location of the work is not within sight of other employees.
- Nonemployees should not work on energized circuits even for the purpose of monitoring or troubleshooting, unless specifically authorized to do so.

SAFETY EQUIPMENT

All employees should be properly trained and wear the required personal protective equipment (PPE) at all times when working with or around electricity.

ADDITIONAL APPAREL CONSIDERATIONS

- Employees are encouraged to wear natural-fiber clothes to prevent material melting.
- Long-sleeved shirts should be worn any time there is danger of arcing.
- No loose, conductive jewelry is to be worn while working on electrical equipment, including necklaces and bracelets.
- Shoes or overshoes should not be relied upon for protection from electrical dangers.
- Contact lenses are not to be worn while working on circuits energized over 120 volts or on any circuit with a capacity over 199 amps.

PERSONAL PROTECTIVE EQUIPMENT

Protective Helmets

- Only protective helmets complying with ANSI Standard Z89.1 at a Class B rating shall be worn on jobs involving electricity. In Ontario, adherence to Canadian Standards Association Standards Z94.1 M1977 complies with the intent of R.R.O. 1980, Reg. 692, S. 84.
- Nothing is to be placed inside the hard hat.

- No holes, for any reason, are to be bored in the shell of the helmet.
- Helmets should not be transported on the rear window shelf of a car or truck.
- Regardless of use, protective helmets should be replaced every five years.

Face and Eye Protection

- Safety glasses with side shields and complying with ANSI Standard Z81.1 must be worn when working on circuits energized at over 50 volts or with a capacity of over 199 amps.
- A faceshield must be worn whenever switching open switches and disconnects and when racking in or out drawout circuit breakers or switches.

RUBBER GOODS

29 CFR § 1910.137 (Electrical Protective Devices) states that all rubber protective equipment for electrical workers shall conform to the following requirements established in the American National Standards Institute Standards (Table 6.2).

Testing and Inspection

- Rubber gloves must be tested by a company-selected testing agency at least every six months; every three months if the gloves are used daily.
- All other rubber goods must be tested at least every twelve months; every six months if they are used at least three days per week.
- All rubber goods are to be inspected and air-tested before each day's use.
- No out-of-date rubber goods will be used.

Storage

- All rubber goods should be stored in an approved container when not in use.
- No rubber goods are to be stored near ozone-producing equipment; this excludes storage near arc-producing equipment.
- If rubber goods must be stored flat, nothing should be stored on top of them.

TABLE 6.2

Item	Standard
Rubber insulating gloves	J6.6-1967
Rubber matting for use around electrical apparatus	J6.7-1935 (R1962)
Rubber insulating blankets	J6.4-1970
Rubber insulating hoods	J6.2-1950 (R1962)
Rubber insulating line hose	J6.1-1950 (R1962)
Rubber insulating sleeves	J6.5-1962

Use

- Rubber gloves must always be used with their leather protectors.
- Rubber gloves must be worn whenever working on energized circuits rated over 300 volts, phase-to-phase or phase-to-ground.
- Rubber sleeves must be used any time hot work over 300 volts is performed inside a deep cabinet or at medium voltages.
- Rubber mats should be used only for secondary shock protection and should not be relied upon as the sole protection against shocks.

FIBERGLASS AND WOOD EQUIPMENT

- Equipment made out of fiberglass and wood (i.e., hot-sticks, switch-sticks, and ladders) must be kept clean and free from grease and oil.
- When equipment becomes scratched or the outer coating begins to chip off, it should be refurbished using a method and material approved by the manufacturer.

ENFORCEMENT OF THE PROGRAM

An electrical safety program without proper enforcement will most likely not meet its potential; therefore, it is recommended that a strong enforcement policy be included with this program. This is to ensure the safety of all employees working with or around electricity.

ELECTRICAL SAFETY CHECKLIST

- Have safety rules been developed for working around electricity?
- Have employees been educated on hazards associated with electricity?
- Are training sessions conducted to train employees on proper work procedures and the use of PPE?
- Are prework discussions held to plan course of duties?
- Is work performed on live voltages?

CONTROL OF HAZARDOUS ENERGY (LOCKOUT AND TAGOUT)

The lack of control of hazardous energy in the industrial workplace is responsible for approximately 10 percent of all serious accidents and a substantial percentage of fatalities in the workplace every year. According to the Bureau of Labor Statistics, failure to shut off power while servicing machinery and other equipment is the primary cause of injuries. To address this hazard, OSHA promulgated its Control of Hazardous Energy standard (commonly referred to as the lockout/tagout standard).

In general, personnel and human resources managers have applauded the lockout/tagout standard. This standard, unlike many preceding standards, is user-friendly and provides the exact sequencing of steps to be followed in order to safeguard employees. The difficulty experienced by many personnel and safety managers is in maintaining compliance with the standard over a period of time.

The most efficient method for safeguarding against accidental activation of machinery is the development of a lockout and tagout program that achieves and maintains compliance with the OSHA standard. By locking out or tagging off power sources, unauthorized use of the machine or equipment is prevented.

A lockout is simply the placement of a substantial locking mechanism on a machine on/off switch or electrical circuit to prevent the power supply from being activated while repairs are made. Lockout procedures are especially effective in preventing injuries to maintenance or repair personnel who may be placed in a hazardous situation by sudden and unexpected activation of machinery while repairs are being made. Lockouts apply not only to electrical hazards but also to all other types of energy (including hydraulic, pneumatic, steam, chemical, and vehicular).

Locking out a machine or piece of equipment can be used in conjunction with, or separate from, the tagging procedures permitted under the standard. Tags are required to be a bright, identifiable color and to be marked with appropriate wording, such as "Danger: Do Not Operate" or other similar warnings. Tags are required to be of a durable nature and are to be affixed securely to all lockout locations. Tags must contain the signature(s) of the employees working on the machinery, the date and time, and the department name or number.

There are four stages in the development and management of an effective lockout and tagout program. The first stage is program development and equipment modification. In this stage, the workplace must be analyzed and a written lockout/tagout program developed. Identification of the potential hazards and, if possible, elimination of exposures to potential injuries by machinery or equipment should be attempted. Additionally, appropriate equipment, including, but not limited to, padlocks, tags, T-bars, and other equipment of a substantial nature, should be purchased during this stage.

The second stage involves the education and training of affected personnel. The personnel and human resources manager must ensure that all training and education are well documented. Hands-on training is highly recommended.

The third stage involves effective monitoring and disciplinary action. The responsibility for ensuring that all employees and equipment covered under this standard are performing in a manner that is in compliance rests solely with the employer. Disciplinary procedures and enforcement thereof are essential to ensuring continued compliance.

The fourth and final stage involves program auditing and program reassessment. The effectiveness of the compliance program can be ensured through periodic evaluation. Program deficiencies can be identified and corrective action initiated to correct the deficiencies identified. The lockout and tagout standard requires employers to establish a written program for locking out and tagging out machinery and equipment. The written program normally contains the following elements:

- Steps for shutting down and securing the machinery; a written program normally details the energy sources for each machine in the workplace and how it should be locked or tagged. All sources of hazardous energy must be listed, and the means for releasing or blocking stored energy should be included.

- Procedural steps for applying locks and tags and specifications for their placement on the equipment or apparatus; identification of the responsible person(s) authorized to apply locks and tags is required.
- Appropriate steps and testing of the machine/equipment required after shutdown and lockout to verify that all energy is safely isolated.
- Procedures to be followed and steps to be taken in restarting the equipment after completion of the work.
- Identification of persons trained and authorized to lockout machinery.
- Procedures for when the task requires group lockout; each employee is required to possess an individual lock, and only the person applying that lock should have a key to the lock. This ensures that, as different team members complete their tasks and remove their locks, remaining members are still fully protected from the hazardous energy.
- Procedures for shift and personnel changes to ensure continuity of the lockout/tagout protection, including provisions for the safe, orderly transfer of lockout/tagout devices between on- and off-duty personnel.
- The requirement that, whenever major replacement, repair, renovation, or modification of machines or equipment is performed and whenever new machines or equipment are installed, energy-isolating devices for such machines or equipment must accept lockout devices.

PREPARING FOR A LOCKOUT OR TAGOUT EVENT

The following procedures must be followed when preparing for a lockout or tagout procedure:

- Conduct a survey to locate and identify all energy-isolating devices to be certain that switches, valves, or other energy-isolating devices apply to the equipment to be locked or tagged out.
- Ensure that all energy sources to a specific piece of equipment have been identified. Machines usually have more than one energy source (electrical, mechanical, or other).
- It is highly recommended to maintain a list of each piece of equipment, with the results of the survey included, so that each time the lockout/tagout procedure must be performed the mechanic or operator can consult the list.

SEQUENCE OF LOCKOUT OR TAGOUT SYSTEM PROCEDURE

These steps should be followed when a machine is locked or tagged:

1. Notify all affected employees that a lockout or tagout system is going to be utilized and the reason for doing so. The authorized employee must know the type and magnitude of energy that the machine or equipment uses and must understand its hazards.
2. If the machine or equipment is operating, shut it down by the normal stopping procedures (depress stop button, open toggle switch, etc.).

3. Operate the switch, valve, or other energy-isolating devices so that the equipment is isolated from its energy sources. Stored energy (such as that in springs, elevated machine members, rotating flywheels, and hydraulic systems or air, gas, steam, water pressure, etc.) must be dissipated or restrained by methods such as repositioning, blocking, or bleeding down.
4. Lockout or tagout the energy-isolating devices with assigned individual locks or tags.
5. After ensuring that no personnel are exposed, and as a check on having disconnected the energy sources, operate the push button or other normal operating controls to make certain the equipment will not operate.
6. The equipment is now locked out or tagged out.

RESTORING MACHINES AND EQUIPMENT TO NORMAL PRODUCTION OPERATIONS

These procedures should be followed to restore machines to normal operation (basically, the reverse of the above lockout section):

1. After the servicing or maintenance is complete and the equipment is ready for normal production operations, check the area around the machines or equipment to ensure that no one is exposed.
2. After all tools have been removed from the machine or equipment, guards have been reinstalled, and employees are in the clear, remove all lockout or tagout devices.
3. Operate the energy-isolating devices to restore energy to the machine or equipment.

TRAINING

Effective employee training and education regarding the lockout/tagout program are vital components to achieving compliance with the OSHA standard:

1. All employees must understand the purpose and use of an energy-control procedure and know what a tagout signifies. They must know why a machine is locked or tagged out and what to do when they encounter a tag or lock on a switch or a device they wish to operate. Because any employee may encounter a lockout tag or lockout, everyone must have a general understanding of lockout/tagout safety.
2. Before machinery shutdown, the authorized employee must know the type and magnitude of energy to be isolated and how to control it. Each machine or type of machine should have a written lockout procedure (preferably attached to the machine).
3. Retraining to ensure employee proficiency must take place when an employee is reassigned to a different area or machine or when written procedures change. Additionally, all new employees must be properly trained regarding the lockout/tagout program. When outside contractors

are brought on site, they should also be informed of the company's lockout procedures. The company, in turn, should be aware of the contractor's procedures and ensure that its personnel understand and comply with the outside contractor's energy-control procedures (as long as they comply with the standard).

4. Each employee authorized to perform maintenance should be fully knowledgeable about all hazardous energy related to specific machinery.
5. The proper sequence of locking out should be fully understood.

When a tagout system is used, employees must be made aware of the following limitations of tags:

- Tags are essentially warning devices affixed to energy-isolating devices; they do not provide the physical restraint on those devices that locks do.
- When a tag is attached to an energy-isolating device, it is not to be removed without authorization of the person responsible for it, and a tag is never to be bypassed, ignored, or otherwise defeated.
- To be effective, tags must be legible and understandable by all authorized employees, all affected employees, and all other employees whose work operations are or may be in the area.
- Tags and their means of attachment must be made of materials able to withstand environmental conditions encountered in the workplace.
- Tags may evoke a false sense of security; their meaning must be understood as part of the overall energy-control program.
- Tags must be securely attached to energy-isolating devices so that they cannot be inadvertently or accidentally detached during use.

A violation of the above OSHA standards poses a substantial risk of great bodily harm or death. Exposure to hazardous chemicals, infected bodily fluids, or harmful energy by employees without the protection afforded under the OSHA standards places the personnel and human resources manager and other company officers in a position of risk for potential civil or criminal sanctions under the OSH Act or state laws in the event of an accident. When developing and managing compliance programs, personnel and human resources managers are advised to ensure compliance with each and every element prescribed in the OSHA standard, completely document compliance with each element of the standard, and properly discipline trained employees who are not complying with the prescribed compliance procedures. Personnel and human resources managers who are unsure as to the requirements of the above standards or any other applicable standard should obtain assistance in order to ensure compliance.

In summation, working with electricity, in any form, is a dangerous situation. Electricity should be handled with care and all appropriate precautions taken at all times. Electricity, or any energy source, cannot tell the difference between individuals or equipment. Electricity is one of our greatest assets, allowing us to accomplish our work; however, it can also be an indiscriminate killer. Proper preparation, training, and precautions can keep electricity in its cage and make it work for you.

However, if you do not properly prepare, this potential killer can be instantaneously released, causing severe damage and death. Be prepared ... electricity is nothing to play with!

Review Questions

1. What consensus and regulatory standards are important considerations when protecting workers from electrical hazards?
2. What are the two main categories of electrical incidents/injuries associated with electrical exposure in the workplace?
3. What are OSHA's fundamental requirements for lockout/tagout?
4. Before working on a live or energized circuit, employees should be able to demonstrate what?

7 Machine Guarding Hazards

Each year, incidents involving human contact with moving machine components result in serious injuries and deaths. Machine guarding and related machinery violations continuously rank among the top ten OSHA citations issued. In fact, the Mechanical Power Transmission (29 C.F.R. 1910.219) and Machine Guarding: General Requirements (29 C.F.R. 1910.212) standards accounted for the sixth and seventh top OSHA violations for FY 1997, with 3077 and 3050 federal citations issued, respectively. There are as many hazards created by moving machine parts as there are types of machines. Safeguards are essential for protecting workers from needless and preventable injuries. A good rule to remember is that any machine part, function, or process that may cause injury must be safeguarded. Three general areas that must be considered for guarding include

1. Point of operation
2. Power transmission areas
3. Other related moving parts

For point of operation, the basic machine operation considered is the purpose for which the machine was designed. It may be sawing, drilling, bending, shearing, punching, or otherwise modifying the original material to make it more useful. These same operations often require tremendous force to transform the material. Machine guards must be in place to prevent contact with these hazardous contact points. Often, loose clothing can become entangled with rotating parts and pull the victim into the machine before it can be stopped. Lacerations, amputations, and crushing injuries may result, the severity of which can range from minor injuries to fatalities. The need to facilitate different kinds, shapes, and sizes of material sometimes makes it difficult to install suitable guards at the point of operation, in which case other means for isolating a worker from this hazard area must be employed. Light curtains, interference fences, and electric interlocks, as well as techniques that physically separate the worker from the machine or prevent machine operation if the worker is present, can be successful if they are well designed, installed, and maintained.

Hazards associated with the power transmission areas of machines may be close to the point of operation or may be located at a distance from it. To gain mechanical advantage, many machines employ pulleys, gears, cams, and levers to increase or decrease speed or torque, to facilitate movement of cutting heads, or to feed the stock to the point of operation. All of these areas have the potential for crushing or amputating body parts if guards are not in place to prevent access to the components. If these parts are not fully enclosed on the machine, they must have suitable guards to prevent contact. Even totally enclosed machines can cause serious injury if maintenance personnel do not lock out all energy sources before opening up the machine to perform maintenance.

Finally, many machines have other related equipment that is used to position the material being processed before, during, or after the point of operation. Examples include pre- and post-positioning tables, feed rollers, material-handling belts, rollers, robot arms, and so on. When a worker steps into, or places a body part in, the path of these machine components, flesh and bone are usually no match for the strength of the machine.

The basic types of hazardous mechanical motions include

1. Rotating
2. Reciprocating
3. Transversing

Rotating motion (including in-running nip points) is particularly dangerous because it can quickly pull fingers, arms, and bodies into the machine before the person can escape or stop the machine. Any projection on a rotating surface increases the likelihood that even slow-turning shafts can pull a victim into the machine. In many cases, the rotating part may be the saw or drill designed to perform the desired operation. Even machine operators who are aware of the danger and have a long history of successful operation can be injured or killed in a single, careless instant. The obvious solution is to design the machine–human interface in a manner that makes it impossible for the machine to operate if all guards are not in place or if a worker is in the immediate area.

Reciprocating motion is another common type of machine motion that transforms linear motion into rotating motion or vice versa. The crankshaft is a common example. It transforms the downward power stroke of the piston into a rotating motion. Many types of saws use reciprocating motion to cut in one direction or stroke and to clear the cutting chips from the work on the opposite stroke. This back-and-forth action can be horizontal or vertical. Although this type of motion is less likely to pull a person into the machine, it is still possible. If a shirtsleeve is caught on the backstroke of a reciprocating saw, the person's arm may be pulled into place to be cut on the next forward stroke. In this type of motion, the cutting tool is either pushed or pulled across or through the surface of the material being processed or the tool is held rigid and the material is moved back and forth. In addition to the hazard posed by the cutting blade(s), a worker could be injured by the moving machine components or the work material. If the machine moves back and forth at a high rate of speed, a worker could be struck many times in succession before escaping the path.

The third type of machine motion is transversing, which involves movement in a straight, continuous line. A belt or chain uses this principle to turn adjacent shafts or carry the material along an assembly line. Any stationery support or equipment along the path of the transverse motion is a potential nip or shear point. The intersection of belts and chains with their gears and pulleys are run-in or nip points. Murphy's Law, which proposes that anything that can go wrong will, suggests that if it is possible for a finger or larger body part to physically fit in these spaces, then eventually they will if left unguarded.

FORMING PROCESS ACTIONS

CUTTING

Cutting operations can involve any of the three basic machine operations above. The principal danger is at the point of operation of the cutting tool(s). Hand saws, band saws, milling machines, and drill presses can injure workers who come in contact with the tools or materials ejected from the work surface in the form of flying chips. These cutting chips can strike anywhere on the body, but the eyes, face, head, and skin are most often injured. Administrative controls, engineering controls, and personal protective equipment can be used to prevent and minimize injuries.

PUNCHING

Punching operations employ the use of power and force to transform material into different shapes, to cut holes in the material, or to trim excess material. A single motion of a press may turn a blank sheet of material into several finished pieces. In other operations, the material may move from one press to another until the finished part emerges from the last press. The size of the hazard increases with the size of the parts processed and the size of the presses. The automobile manufacturing industry uses huge hydraulic presses to transform flat sheet metal into body panels. A careless person who has managed to bypass safety devices and comes in contact with the mating halves of a die during its operation would probably lose whatever body parts the die closed on. This is not meant to indicate that small presses are any less dangerous. Even a tiny press may exert several hundred pounds of pressure. Fingers, hands, and arms can be crushed or amputated by even small presses. Installation or removal of dies during changeover or maintenance are other opportunities for injury. One half of a large press die may weigh several hundred pounds.

SHEARING

Shearing operations are similar to punch-press operations except that the dies are replaced with blades that cut the material passing through. Depending upon the material being cut, this operation may involve hundreds of pounds of pressure. Even a paper shear can easily amputate fingers or arms. Such shears can be driven by pneumatic, hydraulic, or even human power, with the use of simple machines such as the lever and cam.

BENDING

Bending involves the use of presses, breaks, or tubing benders to reshape the material. The primary difference between punching and bending is that bending does not remove material from the work piece. For example, aluminum-siding contractors transform flat aluminum trim stock into various complex shapes up to 12 feet long using a common aluminum break. Any stamping machine that employs dies to reshape the part is actually bending the part. Another example is a tubing bender that transforms straight tubing into complex shapes such as an automobile exhaust pipe. Computerized tubing benders are loaded with an appropriate length of straight tube. When activated, the computer advances and turns the stock as needed to produce the exact shape for which it was programmed. An employee too close to the operation could be struck by the stock or the tubing-bender components. Light curtains, interlocks, personal protective equipment, and education are the keys to safety with bending machines.

WHAT HAPPENS WHEN THE SAFETY DEVICES FAIL?

Each type of safety device employed is a human-designed and developed piece of equipment that is subject to wear and failure. Safety professionals must analyze the impact of various failure modes on the safe operation of the machine. For example, if a light curtain is wired in such a manner that the photocell receiver completes the circuit to the ground when the beam is received, what happens if the receiver is shorted to the ground by a frayed wire? Will the safety device act as if the path is clear even though a human hand may be in the work area?

Failure mode and effect analysis (FMEA) is an important hazard analysis that is too often underutilized. Fail-safe is the proper way to design and wire safety devices. Employees anxious to operate machines with less effort or at higher production rates can be extremely innovative in finding ways to circumvent safety devices. An astounded safety manager asked an employee adjusting a light curtain from behind the device (between the press dies) if the machine was properly locked out. The employee responded, "No, do you think I'm stupid enough to cycle the machine while I'm in it?" Education is the only way to overcome these issues of incorrectly compensated risk.

Frontline supervisors interact with the employees more than anyone else in the plant does and stand the best chance of ensuring that the employees do not take unnecessary risks. Therefore, one of the best ways to improve worker safety is to hold supervisors accountable for the safety of their workers and to make employee safety performance a significant portion of a supervisor's performance appraisals. Ideally, the supervisor would be positively reinforced with incentives if the workers exhibited safe work practices. These same incentives could be withheld if the supervisor failed to take this portion of the job seriously and allowed unsafe work practices to continue.

MACHINE/EQUIPMENT MAINTENANCE

Any time maintenance is being performed on a piece of equipment, the machine should be locked out in accordance with 29 C.F.R. 1910.147. Proper procedures for shutdown, lockout, and startup must be followed to prevent injury. Fixed guards should be designed to allow simple maintenance, such as lubrication, to be performed with the guards in place. Point-of-operation guards may have to be designed based upon the size of the work piece and the nature of the process. They should be sturdy and made of noncombustible material.

For discussion purposes, we can classify safeguards as follows:

1. Types of guards
   a. Fixed
   b. Interlocked
   c. Adjustable
   d. Self-adjusting
2. Guard-related devices
   a. Presence-sensing detectors
      i. Photoelectrical (optical)
      ii. Radiofrequency (capacitance)
      iii. Electromechanical
   b. Pullback
   c. Restraint
   d. Safety controls
      i. Safety trip control
         − Pressure-sensitive body bar
         − Safety tripod
         − Safety tripwire cable
      ii. Two-hand control
      iii. Two-hand trip
   e. Isolation gates
      i. Interlocked
      ii. Other
3. Location/distance from hazard
4. Material-handling methods to improve safety for the machine operator
   a. Automatic feed
   b. Semi-automatic feed
   c. Automatic ejection
   d. Semi-automatic ejection
   e. Robot
5. Miscellaneous aids
   a. Awareness barriers
   b. Miscellaneous protective shields
   c. Hand-feeding tools and holding fixtures

An example of a machine-guarding checklist is located in Appendix H. This checklist is taken from OSHA publication #3067, Concepts and Techniques of Machine Guarding.

Review Questions

1. What are the three general areas that must be considered when evaluating hazards associated with machinery and guarding?
2. What is the difference between a cutting operation and a shearing operation?
3. What should happen during a machine guarding analysis with respect to the reliability of machine guards?
4. What are the four main categories of machine guards and how do they protect employees?

8 Fall Hazards

Deaths from falls top OSHA's "Fatal Four" list for construction. In fact, 364 out of 937 total construction deaths in 2015 (or 38.8 percent of all 2015 construction deaths) were caused by falls. Hands down, falls are the perennial, leading cause of death and serious injury in construction. While not as prevalent as seen in construction, falls still account for a significant number of deaths and serious injuries in general industry. Investigation of these incidents reveals several contributing factors, including:

- Unstable working surfaces
- Misuse of fall-protection equipment
- Human error

Effective countermeasures that have proven successful include the following:

- Guardrails
- Fall-arrest systems
- Safety nets
- Covers
- Travel-limiting systems

In November of 2016, just prior to the publication of this second edition, OSHA issued a revised rule on fall protection and walking–working surfaces for general industry. In addition to effecting common general industry employers, it also impacts building management services, utilities, warehousing, retail, window cleaning, chimney sweeping, and outdoor advertising. Within these industries the new rule addresses floors, stairs, roofs, ladders, ramps, scaffolds, elevated walkways, fall-protection systems, and rope descent systems (RDS). Key compliance dates and deadlines for this rule are detailed below in Table 8.1.

Key themes in the new rule with respect to fall protection in general industry are

1. Become more performance oriented with respect to choices in means of fall protection
2. Establish trigger height for fall protection that is more consistent with construction
3. Establish performance criteria for different types of fall-protection systems

Performance criteria and practices for fall protection means such as guardrails, handrails, stair rails, safety nets, ladder safety systems, and so on, are covered in 29 CFR § 1910.29. Performance criteria for personal fall protection/arrest systems are covered in 29 CFR 1910.140.

TABLE 8.1

Final Subpart D Section and Requirement	Compliance Date
29 CFR § 1910.27(b)(1)—Certification of anchorages.	November 20, 2017
29 CFR § 1910.28(b)(9)(i)(A)—Deadline by which employers must equip existing fixed ladders with a cage, well, ladder safety system, or personal fall-arrest system.	November 19, 2018
29 CFR § 1910.28(b)(9)(i)(B)—Deadline by which employers must begin equipping new fixed ladders with a ladder safety system or personal fall-arrest system.	November 19, 2018
29 CFR § 1910.28(b)(9)(i)(D)—Deadline by which all fixed ladders must be equipped with a ladder safety system or fall arrest system.	November 18, 2036
29 CFR § 1910.36 (a) and (b)—Deadline by which employers must train employees on fall and equipment hazards.	May 17, 2017

The following examples demonstrate trigger heights and fall protection choices for various circumstances/hazards under the new general industry rule:

- Unprotected Sides and Edges
    - Trigger height: Four feet
    - Choice of methods for fall protection:
        – Guardrail systems
        – Safety net systems
        – Personal fall-protection systems
- Hoist Areas (any elevated access opening to a walking–working surface through which equipment or materials are loaded or received)
    - Trigger height: Less than four feet from lower level
    - Choice of methods for fall protection:
        – Guardrail systems
        – Personal fall-arrest system (PFAS)
        – Travel restraint system
- Holes (a gap or open space in a floor, roof, horizontal walking–working surface, or similar surface that is at least two inches [five centimeters] in its least dimension)
    - Trigger height: Less than four feet from lower level
    - Choice of methods for fall protection:
        – Covers
        – Guardrail systems
        – PFAS
        – Travel restraint system
- Dockboards (a portable or fixed device that spans a gap or compensates for a difference in elevation between a loading platform and a transport vehicle.

Dockboards include, but are not limited to, bridge plates, dock plates, and dock levelers)

- Trigger height: Less than four feet from lower level
- Choice of methods for fall protection:
  - Guardrail systems
  - Handrails
- Runway or Other Similar Walkways (*runway* means an elevated walking–working surface, such as a catwalk, a foot walk along shafting, or an elevated walkway between buildings)
  - Trigger height: Less than four feet from lower level
  - Choice of methods for fall protection
    - Guardrail systems
  - Infeasibility exception
  - Applies to having guardrails on *both* sides of runway, *and* runway is used *exclusively* for special purpose
  - Omit guardrail on one side provided that
    - Runway is at least 18 inches wide
    - Each employee is provided with and uses PFAS or travel restraint system
- Protection from openings (a gap or open space in a wall, partition, vertical walking–working surface, or similar surface that is at least 30 inches [76 centimeters] high and at least 18 inches [46 centimeters] wide, through which an employee can fall to a lower level)
  - Trigger heights:
    - Inside bottom edge is less than 39 inches above the walking–working surface *and*
    - Outside bottom edge of opening is less than four feet above lower level
  - Fall protection options:
    - Guardrail systems
    - Safety net systems
    - Travel restraint systems
    - PFAS

These are just some of the circumstances and fall protection options provided in the new rule. The reader should additionally consult the standards to learn about exceptions and special circumstances. 29 CFR §§ 1910.28 is provided in the appendices for reference.

The new rule also addresses protection on fixed ladders. For fixed ladders that extend greater than 24 feet above a lower level, employers must ensure a PFAS, ladder safety system, cage, or well for fixed ladders installed prior to November 19, 2018. On or after November 19, 2018, PFAS or ladder safety systems must be installed on fixed ladders. Replacement ladders should be equipped with a PFAS or ladder safety system. Finally, all fixed ladders must have a PFAS or ladder safety system installed by November 18, 2036.

When discussing some key general industry criteria and practices for fall-protection systems and falling object protection, there are some key definitions that must be understood. Select key definitions include

- *Competent person* means a person who is capable of identifying existing and predictable hazards in any personal fall-protection system or any component of it, as well as in their application and uses with related equipment, and who has authorization to take prompt, corrective action to eliminate the identified hazards.
- *Qualified* describes a person who, by possession of a recognized degree, certificate, or professional standing, or who by extensive knowledge, training, and experience has successfully demonstrated the ability to solve or resolve problems relating to the subject matter, work, or project.
- *Personal fall-protection system* means a system (including all components) an employer uses to provide protection from falling or to safely arrest an employee's fall if one occurs. Examples of personal fall-protection systems include personal fall-arrest systems, positioning systems, and travel restraint systems.
- *Personal fall-arrest system* means a system used to arrest an employee in a fall from a walking–working surface. It consists of a body harness, anchorage, and connector. The means of connection may include a lanyard, deceleration device, lifeline, or a suitable combination of these.
- *Positioning system (work-positioning system)* means a system of equipment and connectors that, when used with a body harness or body belt, allows an employee to be supported on an elevated vertical surface, such as a wall or window sill, and work with both hands free. Positioning systems also are called *positioning system devices* and *work-positioning equipment*.
- *Travel restraint system* means a combination of an anchorage, anchorage connector, lanyard (or other means of connection), and body support that an employer uses to eliminate the possibility of an employee going over the edge of a walking–working surface.

When evaluating a personal fall-protection system, the safety professional must be diligent in her or his research and analysis. There are twenty-two general requirements for personal fall-protection systems, and five additional performance criteria and two usage criteria for PFASs. Body belts are prohibited as part of PFAS, but may be used as part of a positioning system. Again the safety professional must be mindful that there are four additional system performance criteria for positioning systems.

The new general industry rule also mandates inspections and training with respect to fall protection. Employers must ensure that inspections occur regularly and as necessary and hazardous conditions found in those inspections must be corrected or repaired before an employee uses the walking–working surface again. If the correction/repair cannot be made immediately, the hazard must be guarded. If a repair involves the structural integrity of the walking–working surface, a qualified person must perform or supervise the correction/repair.

With respect to training, the new rule requires that before any employee is exposed to a fall hazard, the employer must provide training for each employee who uses personal fall-protection systems or who is required to be trained as specified elsewhere in this subpart. Employers must ensure employees are trained in the requirements of 29 CFR § 1910.30 on or before May 17, 2017. Additionally, the rule requires that the employer ensure that each employee is trained by a qualified person. Training topics specified in the standard include

- The nature of the fall hazards in the work area and how to recognize them.
- The procedures to be followed to minimize those hazards.
- The correct procedures for installing, inspecting, operating, maintaining, and disassembling the personal fall-protection systems that the employee uses.
- The correct use of personal fall-protection systems and equipment specified in paragraph (a)(1) of 29 CFR § 1910.30, including, but not limited to, proper hook-up, anchoring, and tie-off techniques, and methods of equipment inspection and storage, as specified by the manufacturer.
- The proper care, inspection, storage, and use of equipment covered by this subpart before an employee uses the equipment.
- Other specific topics dealing with dockboards, rope descent systems, and so on.

The employer must retrain an employee when the employer has reason to believe the employee does not have the understanding and skill required by paragraphs (a) and (b) of this section. Situations requiring retraining include, but are not limited to, the following:

- When changes in the workplace render previous training obsolete or inadequate.
- When changes in the types of fall-protection systems or equipment to be used render previous training obsolete or inadequate.
- When inadequacies in an affected employee's knowledge or use of fall-protection systems or equipment indicate that the employee no longer has the requisite understanding or skill necessary to use equipment or perform the job safely.

Finally, the rule requires that the training be understandable. Providing understandable training means that the employer must provide information to each employee in a manner that the employee understands.

In 29 C.F.R. 1926.500, OSHA utilizes the following definitions concerning fall protection and related equipment:

- Anchorage: Secure point of attachment for lifelines, lanyards, or deceleration devices.
- Body belt (safety belt): A strap with means both for securing it about the waist and for attaching it to a lanyard, lifeline, or deceleration device.

- Body harness: Straps that may be secured about the employee in a manner that will distribute the fall-arrest forces over at least the thighs, pelvis, waist, chest, and shoulders, with means for attaching it to other components of a personal fall-arrest system.
- Buckle: Any device for holding the body belt or body harness closed around the employee's body.
- Connector: Device used to couple (connect) parts of the personal fall-arrest system and positioning-device systems together. It may be an independent component of the system, such as a carabiner, or it may be an integral component of part of the system, such as a buckle or D-ring sewn into a body belt or body harness or a snap-hook spliced or sewn to a lanyard or self-retracting lanyard.
- Controlled access zone (CAZ): Area in which certain work (e.g., overhand bricklaying) may take place without use of guardrail systems, personal fall-arrest systems, or safety net systems, and access to the zone is controlled.
- Dangerous equipment: Equipment (such as pickling or galvanizing tanks, degreasing units, machinery, electrical equipment, and other units) that, as a result of form or function, may be hazardous to employees who fall onto or into such equipment.
- Deceleration device: Any mechanism, such as a rope grab, rip-stitch lanyard, specially woven lanyard, tearing or deforming lanyards, automatic self-retracting lifelines/lanyards, and so on, that serves to dissipate a substantial amount of energy during a fall arrest or otherwise limit the energy imposed on an employee during fall arrest.
- Deceleration distance: Additional vertical distance a falling employee travels, excluding lifeline elongation and free-fall distance before stopping, from the point at which the deceleration device begins to operate. It is measured as the distance between the location of an employee's body belt or body harness attachment point at the moment of activation (at the onset of fall-arrest forces) of the deceleration device during a fall, and the location of that attachment point after the employee comes to a full stop.
- Equivalent: Alternative designs, materials, or methods to protect against a hazard which the employer can demonstrate will provide an equal or greater degree of safety for employees than the methods, materials, or designs specified in the standard.
- Failure: Load refusal, breakage, or separation of component parts. Load refusal is the point at which the ultimate strength is exceeded.
- Free fall: Act of falling before a personal fall-arrest system begins to apply force to arrest the fall.
- Free fall distance: Vertical displacement of the fall-arrest attachment point on the employee's body belt or body harness between onset of the fall and just before the system begins to apply force to arrest the fall. This distance excludes deceleration distance and lifeline/lanyard elongation, but includes

any deceleration device slide distance or self-retracting lifeline/lanyard extension before they operate and fall-arrest forces occur.

- Guardrail system: Barrier erected to prevent employees from falling to lower levels.
- Hole: Gap or void two inches (5.1 centimeters) or more in its least dimension, in a floor, roof, or other walking–working surface.
- Infeasible: Being impossible to perform the construction work using a conventional fall-protection system (i.e., guardrail system, safety-net system, or personal fall-arrest system) or it is technologically impossible to use any one of these systems to provide fall protection.
- Lanyard: Flexible line of rope, wire rope, or strap that generally has a connector at each end for connecting the body belt or body harness to a deceleration device, lifeline, or anchorage.
- Leading edge: Edge of a floor, roof, or formwork for a floor or other walking–working surface (such as a deck) that changes location as additional floor, roof, decking, or formwork sections are placed, formed, or constructed. A leading edge is considered to be an unprotected side and edge during periods when it is not actively and continuously under construction.
- Lifeline: Component consisting of a flexible line for connection to an anchorage at one end to hang vertically (vertical lifeline) or for connection to anchorages at both ends to stretch horizontally (horizontal lifeline) and which serves as a means for connecting other components of a personal fall-arrest system to the anchorage.
- Low-slope roof: A roof having a slope less than or equal to 4 in 12 (vertical to horizontal).
- Lower levels: Those areas or surfaces to which an employee can fall. Such areas or surfaces include, but are not limited to, ground levels, floors, platforms, ramps, runways, excavations, pits, tanks, material, water, equipment, structures, or portions thereof.
- Mechanical equipment: All motor- or human-propelled wheeled equipment used for roofing work, except wheelbarrows and mop carts.
- Opening: Gap or void 30 inches (76 centimeters) or more high and 18 inches (48 centimeters) or more wide in a wall or partition through which employees can fall to a lower level.
- Overhand bricklaying and related work: Process of laying bricks and masonry units such that the surface of the wall to be jointed is on the opposite side of the wall from the mason, requiring the mason to lean over the wall to complete the work. Related work includes mason tending and electrical installation incorporated into the brick wall during the overhand bricklaying process.
- Personal fall-arrest system: System used to arrest an employee in a fall from a working level. It consists of an anchorage, connectors, and a body belt or body harness and may include a lanyard, deceleration device, lifeline, or suitable combinations of these. As of January 1, 1998, the use of a body belt for fall arrest is prohibited.

- Positioning-device system: Body belt or body harness system rigged to allow an employee to be supported on an elevated vertical surface, such as a wall, and to work with both hands free while leaning.
- Rope grab: Deceleration device that travels on a lifeline and automatically, by friction, engages the lifeline and locks so as to arrest the fall of an employee. A rope grab usually employs the principle of inertial locking, cam/level locking, or both.
- Roof: Exterior surface on the top of a building. This does not include floors or formwork that, because a building has not been completed, temporarily become the top surface of a building.
- Roofing work: Hoisting, storage, application, and removal of roofing materials and equipment, including related insulation, sheet metal, and vapor barrier work, but not including the construction of the roof deck.
- Safety-monitoring system: Safety system in which a competent person is responsible for recognizing and warning employees of fall hazards.
- Self-retracting lifeline/lanyard: A deceleration device containing a drum-wound line that can be slowly extracted from, or retracted onto, the drum under slight tension during normal employee movement and which, after onset of a fall, automatically locks the drum and arrests the fall.
- Snaphook: Connector comprised of a hook-shaped member with a normally closed keeper, or similar arrangement, that may be opened to permit the hook to receive an object and, when released, automatically closes to retain the object. Snaphooks are generally one of two types: (1) locking type, with a self-closing, self-locking keeper that remains closed and locked until unlocked and pressed open for connection or disconnection; or (2) non-locking type, with a self-closing keeper that remains closed until pressed open for connection or disconnection. As of January 1, 1998, the use of a non-locking snaphook as part of personal fall-arrest systems and positioning-device systems is prohibited.
- Steep roof: A roof having a slope greater than 4 in 12 (vertical to horizontal).
- Toeboard: Low protective barrier that prevents the fall of materials and equipment to lower levels and provides protection from falls for personnel.
- Unprotected sides and edges: Any side or edge (except at entrances to points of access) of a walking–working surface (e.g., floor, roof, ramp, or runway) where there is no wall or guardrail system at least 39 inches (1 meter) high.
- Walking–working surface: Any surface, whether horizontal or vertical, on which an employee walks or works, including, but not limited to, floors, roofs, ramps, bridges, runways, formwork, and concrete-reinforcing steel, but not including ladders, vehicles, or trailers on which employees must be located in order to perform their job duties.
- Warning-line system: Barrier erected on a roof to warn employees that they are approaching an unprotected roof side or edge; designates area in which roofing work may not take place without the use of a guardrail, body belt, or safety net systems to protect employees in the area.
- Work area: That portion of a walking–working surface where job duties are being performed.

FALL-PROTECTION CATEGORIES

Fall-protection equipment fits into four functional categories:

1. Fall-arresting
2. Positioning
3. Suspension
4. Retrieval

FALL-ARRESTING EQUIPMENT

A fall-arrest system is required if any risk exists that a worker may fall from an elevated position; as a general rule, the fall-arrest system should be used any time a working height of six feet or more is reached. Working height is the distance from the walking–working surface to a grade or lower level. A fall-arrest system only comes into service when a fall occurs. A full-body harness with a shock-absorbing lanyard or a retractable lifeline is the only product recommended. A full-body harness distributes the forces throughout the body, and the shock-absorbing lanyard decreases the total fall-arresting forces.

POSITIONING

This system holds the worker in place while keeping his or her hands free to work. When the worker leans back, the system is activated; however, the personal positioning system is not specifically designed for fall-arrest purposes.

Effective January 1, 1998, body belts are not an acceptable part of a personal fall-arrest system. A body belt is still acceptable as a positioning device and is regulated by 29 C.F.R. 1926.502(e). The dangers posed by using a body belt as a component in a fall-arrest system include the following:

- Falling out of the belt
- Serious internal injuries from deceleration forces
- Asphyxiation through prolonged suspension by the belt

SUSPENSION SYSTEMS AND EQUIPMENT

This equipment lowers and supports the worker, allowing the worker to use both hands. These systems are widely used in the window-washing and painting industries. Suspension system components are not designed to arrest a free fall and should be used in conjunction with a backup fall-arrest system.

RETRIEVAL PLAN

Preplanning for victim retrieval in the event of a fall should be an integral part of the design of a fall-arresting system. This will facilitate quicker access to the victim in the event that medical treatment is required.

FALL-PROTECTION SYSTEMS

Listed below are different types of fall safety equipment and their recommended usage.

- Cable positioning lanyard: Designed for corrosive or excess heat environments; must be used in conjunction with shock-absorbing devices.
- Class 1: Body belts (single or double D-ring) designed to restrain a person in a hazardous work position and to reduce the possibility of falls; they should not be used when fall potential exists, as they are for positioning only.
- Class 2: Chest harnesses used when there are only limited fall hazards (no vertical free-fall hazard) or for retrieving persons, such as removal of persons from a tank or a bin.
- Class 3: Full-body harnesses designed to arrest the most severe free falls; best choice of harness for absorbing deceleration forces.
- Class 4: Suspension belts, which are independent work supports used to suspend a worker, such as boatswain's chairs or raising or lowering harnesses.
- Rail system: Can be used on any fixed ladders as well as curved surfaces as a reliable method of fall prevention.
- Retractable-lifeline system: Provides fall protection and mobility to user when working at a height or in areas where there is a danger of falling.
- Rope grab: Deceleration device that travels on a lifeline and is used to safely ascend or descend ladders or sloped surfaces; automatically, by friction, engages the lifeline and locks to arrest a person's fall.
- Rope lanyard: Offers some elastic properties for all types of arrest; used for restraint purposes.
- Safety net: Can be used to lessen the fall exposure when temporary floors and scaffolds are not used and the fall distance exceeds 25 feet.
- Shock absorbers: When used, the fall-arresting force will be greatly reduced if a fall occurs.
- Web lanyard: Ideal for restraint purposes where fall hazards are less than two feet.

DUTY TO HAVE FALL PROTECTION

In general, 29 C.F.R. 1926.501 requires protection from falling for anyone who could fall six feet or more. Some notable exceptions are found in various sections of 29 C.F.R. 1926, subpart M, including:

- Requirements relating to roofing work on low-slope roofs, 50 feet (15.25 meters) or less in width, allow the use of a safety-monitoring system alone (i.e., without the warning-line system).
- When the employer can demonstrate that it is infeasible or creates a greater hazard to use these systems, the employer shall develop and

implement a fall-protection plan that meets the requirements of 29 C.F.R. 1926.502(k).

- The provisions of 29 C.F.R. 1926, subpart M, do not apply when employees are making an inspection, investigation, or assessment of workplace conditions prior to the actual start of construction work or after all construction work has been completed.
- Requirements relating to fall protection for employees working on scaffolds are provided in subpart L of 29 C.F.R. 1926.
- Requirements relating to fall protection for employees working on certain cranes and derricks are provided in subpart N of 29 C.F.R. 1926.
- Requirements relating to fall protection for employees performing steel erection work are provided in 1926.105 and in subpart R of 29 C.F.R. 1926.
- Requirements relating to fall protection for employees working on certain types of equipment used in tunneling operations are provided in subpart S of 29 C.F.R. 1926.
- Requirements relating to fall protection for employees engaged in the construction of electric transmission and distribution lines and equipment are provided in subpart V of 29 C.F.R. 1926.
- Requirements relating to fall protection for employees working on stairways and ladders are provided in subpart X of 29 C.F.R. 1926.

TRAINING REQUIREMENTS AS SET FORTH IN 29 C.F.R. 1926.503

The employer shall provide a training program for each employee who might be exposed to fall hazards. The program shall enable each employee to recognize the hazards of falling and shall train each employee in the procedures to be followed in order to minimize these hazards. The employer shall ensure that each employee has been trained, as necessary, by a competent person qualified in the following areas:

- Nature of fall hazards in the work area
- Correct procedures for erecting, maintaining, disassembling, and inspecting the fall-protection systems to be used
- Use and operation of guardrail systems, personal fall-arrest systems, safety-net systems, warning-line systems, safety-monitoring systems, controlled-access zones, and other protection to be used
- Role of each employee in the safety-monitoring system when this system is used
- Limitations on the use of mechanical equipment during the performance of roofing work on low-sloped roofs
- Correct procedures for the handling and storage of equipment and materials and the erection of overhead protection
- Role of employees in fall-protection plans
- Standards contained in this subpart

CERTIFICATION OF TRAINING

The employer shall verify compliance with paragraph (a) of this section by preparing a written certification record that contains:

- Name or other identity of the employee trained
- Date(s) of the training
- Signature of person who conducted the training or signature of the employer

If the employer relies on training conducted by another employer or completed prior to the effective date of this section, the certification record shall indicate the date the employer determined the prior training was adequate rather than the date of actual training. The latest training certification shall be maintained.

RETRAINING

When the employer has reason to believe that any affected employee who has already been trained does not have the understanding and skill required by paragraph (a) of this section, the employer shall retrain each such employee. Circumstances where retraining is required include, but are not limited to, situations where

- Changes in the workplace render previous training obsolete.
- Changes in the types of fall-protection systems or equipment to be used render previous training obsolete.
- Inadequacies in an affected employee's knowledge or use of fall-protection systems or equipment indicate that the employee has not retained the requisite understanding or skill.

9 Occupational Noise Hazards

In many companies and operations, machinery and activities produce excessive occupational noise exposure, which can have a permanent impact on the employees working in and around the area over a period of time. In order to protect American workers from noise-induced, occupational hearing loss, OSHA promulgated the regulation on Hearing Conservation, 29 CFR § 1910.95. In short, employers exposing employees to noise levels at or above the action level (85 dBA on a time-weighted average) or the permissible exposure limit (90 dBA on a time-weighted average) must take specific actions to protect employees' hearing and comply with this standard. Employers whose work environments do not expose employees above the action level are not required to comply with the standard.

As noted above, exposure to excessive occupational noise over a period of time can cause substantial and permanent hearing loss. The OSHA standard specifies that, where the sound level exceeds an eight-hour, time-weighted average (TWA) level of 85 decibels, adherence to the standard is required. If required to do so, most companies address the potential of engineering measures to reduce the noise levels where possible in the work area before addressing the administrative measures identified in the standard. Where engineering controls cannot reduce the noise levels below 85 decibels, companies are responsible for strict adherence to the OSHA standard, including such requirements as audiometric testing, personal protective equipment, training posting, and other requirements as identified in the standard below.

One of the most important components of any hearing conservation program is training. With this standard, individual employees who are required to wear hearing protection must be appropriately trained to wear and care for this important equipment. Training required under the Hearing Conservation standard must include the following topics:

- Effect of noise on hearing
- Purpose of hearing protectors
- Advantages, disadvantages, and attenuation rates of various types of hearing protection
- Hearing protection selection, fitting, use, and care
- Purpose of audiometric testing, with an explanation of testing procedures

This important training is usually performed by the safety manager, training officer, or other professional who has acquired the appropriate education. Although supervisors do not normally prescribe the types of protection necessary or provide fitting instructions for hearing conservation protection (i.e., earplugs), it is important that supervisors inspect their employees on a daily basis to ensure that they are properly wearing and caring for their hearing protection. Inspection and monitoring of the wearing of hearing protection by employees can be incorporated into a supervisor's daily safety inspections or can be performed as a separate inspection.

An often-overlooked area that is not addressed in this standard is the disposal of the hearing protection after use. Employees should be trained in, and supervisors should ensure, proper disposal and replacement of hearing protection to prevent finding discarded hearing protection in products or in machinery. Any employee not complying with the requirements of the program should be subject to disciplinary action in accordance with company policy.

Unlike other OSHA standards, a written program for hearing conservation is not required; however, most prudent companies and organizations do develop a written hearing conservation program in order to guide and document their efforts. Additionally, many companies and organizations maintain the required audiometric licensure, training certificates, and calibration documents within their written programs for easy monitoring and access.

In summation, the loss of an individual's hearing can be devastating, and companies with high noise–exposure levels can be hurt by the high cost of claims from long-term employees who have worked in the noise-exposure areas. Prudent companies and organizations should appropriately monitor and assess their work areas to identify potential exposures, if any, and take the required steps to ensure compliance with the OSHA Hearing Conservation standard and to protect their employees from the long-term health effects of noise.

HEARING CONSERVATION*

29 C.F.R. 1910.95(A)

Protection against the effects of noise exposure shall be provided when the sound levels exceed those shown in Table 9.1 when measured on the A scale of a standard sound-level meter at slow response. When noise levels are determined by octave-band analysis, the equivalent A-weighted sound level may be determined as follows (see Figure 9.1).

Equivalent sound level contours. Octave-band sound-pressure levels may be converted to the equivalent A-weighted sound level by plotting them on the graph shown in Figure 9.1 and noting the A-weighted sound level corresponding to the point of highest penetration into the sound level contours. This equivalent A-weighted sound level, which may differ from the actual A-weighted sound level of the noise, is used to determine exposure limits from Table 9.1.

TABLE 9.1
Permissible Noise Exposures

Duration per Day (Hours)	Sound Level (Decibels, Slow Response)
8	90
6	92
4	95
3	97
2	100
1-1/2	102
1	105
1/2	110
1/4 or less	115

Note: When the daily noise exposure is composed of two or more periods
of noise exposure of different levels, their combined effect should be
considered, rather than the individual effect of each. If the sum of the
following fractions: $C(1)/T(1) + C(2)/T(2) + ... + C(n)/T(n)$ exceeds
unity, then the mixed exposure should be considered to exceed the
limit value. $C(n)$ indicates the total time of exposure at a specified
noise level, and $T(n)$ indicates the total time of exposure permitted at
that level. Exposure to impulsive or impact noise should not exceed
140-decibel peak sound-pressure level.

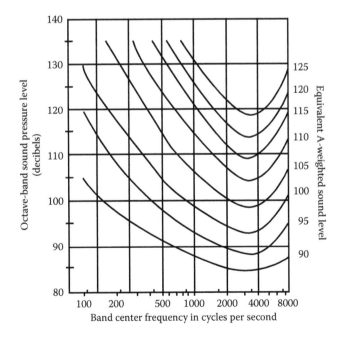

FIGURE 9.1   Equivalent A-weighted sound level.

29 C.F.R. 1910.95(B)

(b)(1) When employees are subjected to sound exceeding those listed in Table 9.1, feasible administrative or engineering controls shall be utilized. If such controls fail to reduce sound levels within the levels of Table 9.1, personal protective equipment shall be provided and used to reduce sound levels within the levels of the table.

(b)(2) If the variations in noise level involve maxima at intervals of one second or less, it is to be considered continuous.

29 C.F.R. 1910.95(C): Hearing Conservation Program

(c)(1) The employer shall administer a continuing, effective, hearing conservation program, as described in paragraphs (c) through (o) of this section, whenever employee noise exposures equal or exceed an eight-hour, time-weighted average sound level of 85 decibels measured on the A scale (slow response) or, equivalently, a dose of 50 percent. For purposes of the hearing conservation program, employee noise exposures shall be computed in accordance with Appendix A and Table 9.1, and without regard to any attenuation provided by the use of personal protective equipment.

(c)(2) For purposes of paragraphs (c) through (n) of this section, an eight-hour, time-weighted average of 85 decibels or a dose of 50 percent shall also be referred to as the *action level*.

29 C.F.R. 1910.95(D): Monitoring

(d)(1) When information indicates that any employee's exposure may equal or exceed an eight-hour, time-weighted average of 85 decibels, the employer shall develop and implement a monitoring program.

    (d)(1)(i) The sampling strategy shall be designed to identify employees for inclusion in the hearing conservation program and to enable the proper selection of hearing protectors.

    (d)(1)(ii) Where circumstances such as high worker mobility, significant variations in sound level, or a significant component of impulse noise make area monitoring generally inappropriate, the employer shall use representative personal sampling to comply with the monitoring requirements of this paragraph unless the employer can show that area sampling produces equivalent results.

(d)(2)

    (d)(2)(i) All continuous, intermittent, and impulsive sound levels from 80 decibels to 130 decibels shall be integrated into the noise measurements.

    (d)(2)(ii) Instruments used to measure employee noise exposure shall be calibrated to ensure measurement accuracy.

(d)(3) Monitoring shall be repeated whenever a change in production, process, equipment, or controls increases noise exposures to the extent that

    (d)(3)(i) Additional employees may be exposed at or above the action level.

(d)(3)(ii) The attenuation provided by hearing protectors being used by employees may be rendered inadequate to meet the requirements of paragraph (j) of this section.

29 C.F.R. 1910.95(E): EMPLOYEE NOTIFICATION

The employer shall notify each employee exposed at or above an eight-hour, time-weighted average of 85 decibels of the results of the monitoring.

29 C.F.R. 1910.95(F): OBSERVATION OF MONITORING

The employer shall provide affected employees or their representatives with an opportunity to observe any noise measurements conducted pursuant to this section.

29 C.F.R. 1910.95(G): AUDIOMETRIC TESTING PROGRAM

(g)(1) The employer shall establish and maintain an audiometric testing program as provided in this paragraph by making audiometric testing available to all employees whose exposures equal or exceed an eight-hour, time-weighted average of 85 decibels.

(g)(2) The program shall be provided at no cost to employees.

(g)(3) Audiometric tests shall be performed by a licensed or certified audiologist, otolaryngologist, or other physician, or by a technician who is certified by the Council of Accreditation in Occupational Hearing Conservation or who has satisfactorily demonstrated competence in administering audiometric examinations, obtaining valid audiograms, and properly using, maintaining, and checking calibration and proper functioning of the audiometers being used. A technician who operates microprocessor audiometers does not need to be certified. A technician who performs audiometric tests must be responsible to an audiologist, otolaryngologist, or physician.

(g)(4) All audiograms obtained pursuant to this section shall meet the requirements of Appendix C: Audiometric Measuring Instruments.

(g)(5) Baseline audiogram:

(g)(5)(i) Within six months of an employee's first exposure at or above the action level, the employer shall establish a valid baseline audiogram against which subsequent audiograms can be compared.

(g)(5)(ii) Mobile test van exception: Where mobile test vans are used to meet the audiometric testing obligation, the employer shall obtain a valid baseline audiogram within one year of an employee's first exposure at or above the action level. Where baseline audiograms are obtained more than six months after the employee's first exposure at or above the action level, employees shall wear hearing protectors for any period exceeding six months after first exposure until the baseline audiogram is obtained.

(g)(5)(iii) Testing to establish a baseline audiogram shall be preceded by at least 14 hours without exposure to workplace noise. Hearing protectors may be used as a substitute for the requirement that baseline audiograms be preceded by 14 hours without exposure to workplace noise.

(g)(5)(iv) The employer shall notify employees of the need to avoid high levels of nonoccupational noise exposure during the 14-hour period immediately preceding the audiometric examination.

(g)(6) Annual audiogram: At least annually after obtaining the baseline audiogram, the employer shall obtain a new audiogram for each employee exposed at or above an eight-hour, time-weighted average of 85 decibels.

(g)(7) Evaluation of audiogram:

(g)(7)(i) Each employee's annual audiogram shall be compared to that employee's baseline audiogram to determine if the audiogram is valid and if a standard threshold shift as defined in paragraph (g)(10) of this section has occurred. This comparison may be done by a technician.

(g)(7)(ii) If the annual audiogram shows that an employee has suffered a standard threshold shift, the employer may obtain a retest within 30 days and consider the results of the retest as the annual audiogram.

(g)(7)(iii) The audiologist, otolaryngologist, or physician shall review problem audiograms and shall determine whether there is a need for further evaluation. The employer shall provide to the person performing this evaluation the following information:

(g)(7)(iii)(A) A copy of the requirements for hearing conservation as set forth in paragraphs (c) through (n) of this section.

(g)(7)(iii)(B) The baseline audiogram and most recent audiogram of the employee to be evaluated.

(g)(7)(iii)(C) Measurements of background sound-pressure levels in the audiometric test room as required in Appendix D: Audiometric Test Rooms.

(g)(7)(iii)(D) Records of audiometer calibrations required by paragraph (h)(5) of this section.

(g)(8) Follow-up procedures:

(g)(8)(i) If a comparison of the annual audiogram to the baseline audiogram indicates that a standard threshold shift as defined in paragraph (g)(10) of this section has occurred, the employee shall be informed of this fact in writing within 21 days of the determination.

(g)(8)(ii) Unless a physician determines that the standard threshold shift is not work-related or aggravated by occupational noise exposure, the employer shall ensure that the following steps are taken when a standard threshold shift occurs:

(g)(8)(ii)(A) Employees not using hearing protectors shall be fitted with hearing protectors, trained in their use and care, and required to use them.

(g)(8)(ii)(B) Employees already using hearing protectors shall be refitted and retrained in the use of hearing protectors and provided with hearing protectors offering greater attenuation if necessary.

(g)(8)(ii)(C) The employee shall be referred for a clinical audiologi-
cal evaluation or an otological examination as appropriate if additional testing is necessary or if the employer suspects that a medical
pathology of the ear is caused or aggravated by the wearing of hearing protectors.

(g)(8)(ii)(D) The employee is informed of the need for an otological
examination if a medical pathology of the ear that is unrelated to
the use of hearing protectors is suspected.

(g)(8)(iii) If subsequent audiometric testing of an employee whose exposure
to noise is less than an eight-hour, time-weighted average of 90 decibels
indicates that a standard threshold shift is not persistent, the employer:

(g)(8)(iii)(A) Shall inform the employee of the new audiometric
interpretation.

(g)(8)(iii)(B) May discontinue the required use of hearing protectors for
that employee.

(g)(9) Revised baseline: An annual audiogram may be substituted for the baseline audiogram when, in the judgment of the audiologist, otolaryngologist,
or physician who is evaluating the audiogram:

(g)(9)(i) The standard threshold shift revealed by the audiogram is persistent.

(g)(9)(ii) The hearing threshold shown in the annual audiogram indicates
significant improvement over the baseline audiogram.

(g)(10) Standard threshold shift:

(g)(10)(i) As used in this section, a standard threshold shift is a change in
hearing threshold relative to the baseline audiogram of an average of
10 decibels or more at 2000, 3000, and 4000 hertz in either ear.

(g)(10)(ii) In determining whether a standard threshold shift has occurred,
allowance may be made for the contribution of aging (presbycusis) to the
change in hearing level by correcting the annual audiogram according
to the procedure described in Appendix F: Calculation and Application
of Age Correction to Audiograms.

29 C.F.R. 1910.95(H): AUDIOMETRIC TEST REQUIREMENTS

(h)(1) Audiometric tests shall be pure-tone, air-conduction, hearing-threshold
examinations with test frequencies including, as a minimum, 500, 1000,
2000, 3000, 4000, and 6000 hertz. Tests at each frequency shall be taken
separately for each ear.

(h)(2) Audiometric tests shall be conducted with audiometers (including microprocessor audiometers) that meet the specifications of and are maintained
and used in accordance with the American National Standard Specification
for Audiometers, S3.6-1969, which is incorporated by reference as specified
in Section 1910.6.

(h)(3) Pulsed-tone and self-recording audiometers, if used, shall meet the
requirements specified in Appendix C: Audiometric Measuring Instruments.

(h)(4) Audiometric examinations shall be administered in a room meeting the requirements listed in Appendix D: Audiometric Test Rooms.

(h)(5) Audiometer calibration:

(h)(5)(i) The functional operation of the audiometer shall be checked before each day's use by testing a person with known, stable hearing thresholds and by listening to the audiometer's output to make sure that the output is free from distorted or unwanted sounds. Deviations of 10 decibels or greater require an acoustic calibration.

(h)(5)(ii) Audiometer calibration shall be checked acoustically at least annually in accordance with Appendix E: Acoustic Calibration of Audiometers. Test frequencies below 500 Hz and above 6000 hertz may be omitted from this check. Deviations of 15 decibels or greater require an exhaustive calibration.

(h)(5)(iii) An exhaustive calibration shall be performed at least every 2 years in accordance with sections 4.1.2, 4.1.3, 4.1.4.3, 4.2, 4.4.1, 4.4.2, 4.4.3, and 4.5 of the American National Standard Specification for Audiometers, S3.6-1969. Test frequencies below 500 hertz and above 6000 hertz may be omitted from this calibration.

29 C.F.R. 1910.95(i): Hearing Protectors

(i)(1) Employers shall make hearing protectors available to all employees exposed to an eight-hour, time-weighted average of 85 decibels or greater at no cost to the employees. Hearing protectors shall be replaced as necessary.

(i)(2) Employers shall ensure that hearing protectors are worn:

(i)(2)(i) By an employee who is required by paragraph (b)(1) of this section to wear personal protective equipment.

(i)(2)(ii) By any employee who is exposed to an eight-hour, time-weighted average of 85 decibels or greater, and who

(i)(2)(ii)(A) Has not yet had a baseline audiogram established pursuant to paragraph (g)(5)(ii) or

(i)(2)(ii)(B) Has experienced a standard threshold shift.

(i)(3) Employees shall be given the opportunity to select their hearing protectors from a variety of suitable hearing protectors provided by the employer.

(i)(4) The employer shall provide training in the use and care of all hearing protectors provided to employees.

(i)(5) The employer shall ensure proper initial fitting and supervise the correct use of all hearing protectors.

29 C.F.R. 1910.95(j): Hearing Protector Attenuation

(j)(1) The employer shall evaluate hearing protector attenuation for the specific noise environments in which the protector will be used. The employer shall use one of the evaluation methods described in Appendix B: Methods for Estimating the Adequacy of Hearing Protection Attenuation.

(j)(2) Hearing protectors must attenuate employee exposure at least to an eight-hour, time-weighted average of 90 decibels as required by paragraph (b) of this section.

(j)(3) For employees who have experienced a standard threshold shift, hearing protectors must attenuate employee exposure to an eight-hour, time-weighted average of 85 decibels or below.

(j)(4) The adequacy of hearing protector attenuation shall be re-evaluated whenever employee noise exposures increase to the extent that the hearing protectors provided may no longer provide adequate attenuation. The employer shall provide more effective hearing protectors where necessary.

29 C.F.R. 1910.95(k): Training Program

(k)(1) The employer shall institute a training program for all employees who are exposed to noise at or above an eight-hour, time-weighted average of 85 decibels and shall ensure employee participation in such program.

(k)(2) The training program shall be repeated annually for each employee included in the hearing conservation program. Information provided in the training program shall be updated to be consistent with changes in protective equipment and work processes.

(k)(3) The employer shall ensure that each employee is informed of the following:

(k)(3)(i) The effects of noise on hearing.

(k)(3)(ii) The purpose of hearing protectors; the advantages, disadvantages, and attenuation of various types; and instructions on selection, fitting, use, and care.

(k)(3)(iii) The purpose of audiometric testing and an explanation of the test procedures.

29 C.F.R. 1910.95(l): Access to Information and Training Materials

(l)(1) The employer shall make available to affected employees or their representatives copies of this standard and shall also post a copy in the workplace.

(l)(2) The employer shall provide to affected employees any informational materials pertaining to the standards that are supplied to the employer by the Assistant Secretary.

(l)(3) The employer shall provide, upon request, all materials related to the employer's training and education program pertaining to this standard to the assistant secretary and the director.

29 C.F.R. 1910.95(m): Recordkeeping

(m)(1) Exposure measurements: The employer shall maintain an accurate record of all employee exposure measurements required by paragraph (d) of this section.

(m)(2) Audiometric tests:

(m)(2)(i) The employer shall retain all employee audiometric test records obtained pursuant to paragraph (g) of this section.

(m)(2)(ii) This record shall include

(m)(2)(ii)(A) Name and job classification of the employee.

(m)(2)(ii)(B) Date of the audiogram.

(m)(2)(ii)(C) The examiner's name.

(m)(2)(ii)(D) Date of the last acoustic or exhaustive calibration of the audiometer.

(m)(2)(ii)(E) Employee's most recent noise exposure assessment.

(m)(2)(ii)(F) The employer shall maintain accurate records of the measurements of the background sound-pressure levels in audiometric test rooms.

(m)(3) Record retention: The employer shall retain records required in this paragraph (m) for at least the following periods:

(m)(3)(i) Noise exposure measurement records shall be retained for two years.

(m)(3)(ii) Audiometric test records shall be retained for the duration of the affected employee's employment.

(m)(4) Access to records: All records required by this section shall be provided upon request to employees, former employees, representatives designated by the individual employee, and the assistant secretary. The provisions of 29 C.F.R. 1910.20 (a)–(e) and (g)–(m)(4)(i) apply to access to records under this section.

(m)(5) Transfer of records: If the employer ceases to do business, the employer shall transfer to the successor employer all records required to be maintained by this section, and the successor employer shall retain them for the remainder of the period prescribed in paragraph (m)(3) of this section.

29 C.F.R. 1910.95(N): Appendices

(n)(1) Appendices A, B, C, D, and E to this section are incorporated as part of this section and the contents of these appendices are mandatory.

(n)(2) Appendices F and G to this section are informational and are not intended to create any additional obligations not otherwise imposed or to detract from any existing obligations.

29 C.F.R. 1910.95(O): Exemptions

Paragraphs (c) through (n) of this section shall not apply to employers engaged in oil and gas well drilling and servicing operations.

29 C.F.R. 1910.95(p): Startup Date

Baseline audiograms required by paragraph (g) of this section shall be completed by March 1, 1984.

[39 FR 23502, June 27, 1974, as amended at 46 FR 4161, Jan. 16, 1981; 46 FR 62845, Dec. 29, 1981; 48 FR 9776, Mar. 8, 1983; 48 FR 29687, June 28, 1983; 54 FR 24333, June 7, 1989; 61 FR 5507, Feb. 13, 1996; 61 FR 9227, March 7, 1996.]

10 Bloodborne Pathogens

OSHA's Bloodborne Pathogen standard was developed to protect employees from possible exposure to HIV, hepatitis, and other bloodborne diseases to which the employee may be exposed in the workplace. For emergency medical services (EMS) organizations, the potential for exposure to bodily fluids (thus, bloodborne pathogens) occurs when medical personnel render treatment in homes, accident scenes, or elsewhere. From a legal perspective, providing emergency medical treatment has numerous potential risks for EMS organizations in regard to not only workers' compensation for exposed employees but also tort liability (i.e., negligence), malpractice, and decreased efficacy.

Compliance with the standard is an absolute must for EMS organizations. The basic procedure for the development of a program to achieve compliance with this new standard includes:

- Acquire a copy of the current OSHA standard.
- Review this standard with the management team and acquire management commitment and appropriate funding for this program.
- Develop a written program incorporating all required elements of the standard which include, but are not limited to, Universal Precautions, Engineering and Work Practice Controls, personal protective equipment, housekeeping, infectious waste disposal, laundry procedures, training requirements, hepatitis B vaccinations, information to be provided to the physician, medical recordkeeping, signs and labels, and availability of medical records.

Under the scope of universal precautions, the company's operations should be analyzed to provide all necessary safeguards to employees who may have possible contact with human blood. Engineering Controls and Work Practice Controls should be examined and evaluated. Employees should be properly trained in the requirements of this standard. Procedures should be established for the safe disposal of used needles, used personal protective equipment, and other equipment. Such areas as the refrigerator, cabinets, or freezers where blood could be stored should be prohibited for the storage of food or drink (e.g., employees keeping their lunches in the nurse's refrigerator). Employees must be advised to refrain from eating, drinking, smoking, applying cosmetics or lip balm, or handling contact lenses after possible exposure.

Where there is the potential of exposure, personal protective equipment, such as surgical gloves, gowns, fluid-proof aprons, faceshields, pocket masks, ventilation devices, and so on must be provided to employees. This requirement is especially important for first-aid responders and plant medical personnel who may be exposed when providing first aid to an injured worker. The personal protective equipment must be appropriately located for easy accessibility. Hypoallergenic personal protective equipment should be made available to employees who may be allergic to the normal personal protective equipment.

Employers are required to maintain a clean and sanitary worksite. A *written* schedule for cleaning and sanitizing (disinfecting) all applicable work areas must be implemented and included in the program. All areas, equipment, and so on that have been exposed (e.g., after an accident) must be cleaned and disinfected. Exposed broken glass may *not* be picked up directly with the hands. A dust broom, vacuum, tongs, or other instrument must be used.

All infectious waste, such as bandages, towels, or other items exposed to human blood, must be placed in a leakproof container or bag bearing the appropriate label. This waste must be properly disposed of as medical waste according to federal, state, and local regulations. Please check with the local hospital or governmental agency to learn about the applicable regulations.

Contaminated uniforms, smocks, and other items of personal clothing must be laundered in accordance with the standard. *Do not send contaminated uniforms, smocks, etc., to the in-plant laundry.*

Where required, employees who are at risk of being exposed must undergo a medical examination and acquire a hepatitis B vaccination. Examples of personnel who may meet this requirement include plant medical personnel and first-aid responders, among others. Training requirements include providing a copy of the OSHA standard and explaining it, as well as discussion of the symptoms and epidemiology of bloodborne diseases; transmission of bloodborne pathogens; methods for recognizing jobs and other activities that could involve exposure to blood; the uses and limitations of practices that will prevent or reduce exposure, such as Engineering Controls and the use of personal protective equipment; the types, proper use, location, removal, handling, decontamination, or disposal of personal protective equipment; the basis for selection of personal protective equipment; the hepatitis B vaccination, including information on its efficacy, safety, and benefits; appropriate actions to be taken and persons to contact in an emergency; procedures to be followed if an exposure incident occurs, including the method of reporting, medical follow-up, and medical counseling; and signs and labels. Additional training may be required in applicable laboratory situations and other circumstances.

The employer is required to maintain complete and accurate medical records. These records must include the names and social security numbers of employees; each employee's hepatitis B vaccination record and medical evaluations; results of all physical examinations, medical testing, and follow-up procedures for each employee; physicians' written opinions; and copies of all information provided to the physicians. The employer is charged with the responsibility of maintaining the confidentiality of these records.

The employer is additionally required to maintain all testing records, which must include the dates of all training; names of persons conducting the training; and names of all participants. The training records must be maintained for a minimum of five (5) years. Employees and OSHA must be given the opportunity to review and copy these records upon request.

All containers of infectious waste (including, but not limited to, refrigerators and freezers containing infectious waste, medical disposal containers, and all other containers) *must* be properly labeled. The required label must be fluorescent orange or orange-red, bear the appropriate symbol, and include the lettering "BIOHAZARD" in a contrasting color. This label *must* be affixed to all containers containing infectious waste and must remain affixed until the waste is properly disposed of.

EXAMPLE OF A BLOODBORNE PATHOGENS PROGRAM

BLOODBORNE PATHOGENS PROGRAM

Developed on _____.   Updated on _____.

Contents	Pages
Introduction	_____
Definitions	_____
Exposure control plan: overview	_____
Exposure determination	_____
Engineering and work practice controls	_____
Personal protective equipment	_____
Housekeeping	_____
Hepatitis B vaccine and vaccination series and postexposure evaluation	_____
Hazard communication	_____
Training and education	_____
Recordkeeping	_____

INTRODUCTION

The objective of this manual is to provide EMS, Inc., employees, health professionals, and other managers who are responsible for protecting the health and safety of employees with information they need to comply with Occupational Safety and Health Administration (OSHA) requirements concerning the elimination or minimization of occupationally transmitted bloodborne infections. Included are instructions for compliance with the OSHA Bloodborne Pathogens standard, education and training materials, and guidelines for the medical evaluation and treatment of employees who may be exposed to blood, body fluids, or other materials potentially contaminated with blood or body fluids. It is worth emphasizing that, without meticulous adherence to the concepts and requirements in this manual, the potential for serious injury to health is a very real one; the inherent complexities attendant with eliminating or minimizing potential exposures should not be underestimated.

Instructions

The manual has been designed to present the Bloodborne Pathogens standard in a readily understandable format and to provide checklists and self-prompting forms that are easy to use and which can assist companies and organizations to achieve compliance. The reader is urged to take a few moments to become familiar with the various sections in this manual. The format of each section consists of text from the standard followed by the recommended forms to be used. Once completed, the forms become the facility's Exposure Control Plan for compliance with the OSHA standard.

Bloodborne Pathogens Program

A copy of the written Bloodborne Pathogens Program and Bloodborne Pathogen Training Program for EMS, Inc., as well as the OSHA standard upon which these programs are based, is located in the main office and is readily available for use by all employees. If you should have any questions, please contact your supervisor.

_____

President

Definitions (OSHA Defined)

Please take a moment to read this section carefully. The words and phrases included have meanings specific to the Bloodborne Pathogens standard.

Blood: Human blood, human blood components, and products made from human blood.

Bloodborne pathogens: Pathogenic microorganisms that are present in human blood and can cause disease in humans. These pathogens include, but are not limited to, hepatitis B virus (HBV) and human immunodeficiency virus (HIV).

Clinical laboratory: A workplace where diagnostic or other screening procedures are performed on blood or other potentially infectious materials.

Contaminated: The presence or reasonably anticipated presence of blood or other potentially infectious materials on an item or surface.

Contaminated laundry: Laundry that has been soiled with blood or other potentially infectious materials or may contain sharps.

Contaminated sharps: Any contaminated object that can penetrate the skin, including, but not limited to, needles, scalpels, broken glass, broken capillary tubes, and exposed ends of dental wires.

Decontamination: The use of physical or chemical means to remove, inactivate, or destroy bloodborne pathogens on a surface or item to the point where they are no longer capable of transmitting infectious particles and the surface or item is rendered safe for handling, use, or disposal.

Employees at significant risk of exposure: Includes healthcare providers (physicians, nurses, lab technicians), who comprise a risk group because they are routinely exposed to body fluids during the performance of their clinical duties. Those employees who are not part of the primary healthcare team (e.g., security personnel, CPR-trained employees) but as part of their job responsibilities are required to respond to medical emergencies may also be at increased risk due to the greater potential for exposure to blood in emergency situations. Employees who handle medical waste, laundry, medical instruments, or other materials contaminated by body fluids may be at risk of exposure during cleaning, maintenance, or other housekeeping activities. Other medical facility staff such as administrators, managers, administrative assistants, receptionists, and so on are generally not at risk.

Engineering controls: Controls (e.g., sharps disposal containers, self-sheathing needles) that isolate or remove the bloodborne pathogens hazard from the workplace.

Exposure incident: Specific eye, mouth, other mucous membrane, nonintact skin, or parenteral contact with blood or other potentially infectious materials that results from the performance of an employee's duties.

Handwashing facilities: A facility providing an adequate supply of running, potable water, soap, and single-use towels or hot-air drying machines.

HBV: Hepatitis B virus.

HIV: Human immunodeficiency virus.

Licensed healthcare professional: A person whose legally permitted scope of practice allows him or her to independently perform the activities required for hepatitis B vaccination and postexposure evaluation and follow-up.

Occupational exposure: Reasonably anticipated skin, eye, mucous membrane, or parenteral contact with blood or other potentially infectious materials that may result from the performance of an employee's duties.

Other potentially infectious materials: (1) The following human body fluids: semen, vaginal secretions, cerebrospinal fluids, synovial fluid, pleural fluid, pericardial fluid, peritoneal fluid, amniotic fluid, saliva in dental procedures, any body fluid that is visibly contaminated with blood, and all body fluids in situations where it is difficult or impossible to differentiate between body fluids; (2) any unfixed tissue or organ (other than intact skin) from a human (living or dead); and (3) HIV-containing cell or tissue cultures, organ cultures, HIV- or HBV-containing culture medium or other solutions, blood organs, or other tissues from experimental animals infected with HIV or HBV.

Parenteral: Piercing mucous membranes or the skin barrier by such events as needle sticks, human bites, cuts, and abrasions.

Personal protective equipment: Specialized clothing or equipment worn by an employee for protection against a hazard. General work clothes (e.g., uniforms, pants, shirts, or blouses) not intended to function as protection against a hazard are not considered to be personal protective equipment.

Regulated waste: Liquid or semi-liquid blood or other potentially infectious materials, contaminated items that would release blood or other potentially infectious materials in a liquid or semi-liquid state if compressed, items that are caked with dried blood or other potentially infectious materials and are capable of releasing these materials during handling, contaminated sharps, and pathological and microbiological wastes containing blood or other potentially infectious materials.

Research laboratory: A laboratory producing or using research laboratory–scale amounts of HIV or HBV. Research laboratories may produce high concentrations of HIV or HBV but not in the volume found in production facilities.

Source individual: Any individual, living or dead, whose blood or other potentially infectious materials may be a source of occupational exposure to the employee. Examples include, but are not limited to, hospital and clinic patients, clients in institutions for the developmentally disabled, trauma victims, clients of drug and alcohol treatment facilities, residents of hospices and nursing homes, human remains, and individuals who donate or sell blood or blood components.

Sterilize: The use of a physical or chemical procedure to destroy all microbial life, including highly resistant bacterial endospores.

Universal precautions: An approach to infection control, according to which all human blood and certain body fluids are treated as if known to be infectious for HIV, HBV, and other bloodborne pathogens.

Work practice controls: Controls that reduce the likelihood of exposure by altering the manner in which a task is performed (e.g., prohibiting recapping of needles by a two-handed technique).

EXPOSURE CONTROL PLAN: OVERVIEW

Each company or organization must have a written Exposure Control Plan (ECP). The purpose of the ECP is to eliminate or minimize *occupational exposures* (hereafter, Exposures). Following are mandatory elements of the ECP which will be addressed in subsequent sections of this manual.

- Exposure determination (identifying covered employees)
- Schedule and methods of implementation:
  - Methods of compliance
  - Hepatitis B vaccination and postexposure evaluation
  - Communication of hazards to employees
  - Recordkeeping
- Procedure for evaluation of circumstances surrounding exposure incidents

Note: Companies or organizations are responsible for ensuring that the ECP is readily available (during the work shift) for review by covered employees and that a copy is made available to them without charge, within *15 working days* upon request. The ECP must be reviewed *annually* and *whenever necessary* to reflect new or modified tasks and procedures which affect occupational exposure, new employee positions with occupational exposure, and changed employee positions which now include occupational exposure.

GENERAL INFORMATION

Facility:	EMS, Inc.
Address:	[street address]
	[city, state, zip]

Bloodborne Pathogens Manual

Completed by: _____

Date: _____   ____

Date(s) of update reviews: _____   ____

Update reviewer(s): _____   ____

Review and updating of the appropriate sections in this *Bloodborne Pathogens Manual* will be performed *annually* and whenever there is a need for a change in the exposure determination list (e.g., change in employees who are occupationally exposed; change in tasks, procedures, or jobs that may affect occupational exposure).

Responsible person/party: _____

A copy of the *Bloodborne Pathogens Manual* will be made readily accessible to all occupationally exposed employees at [location]. Provisions have been made to provide a copy of the Exposure Control Plan within 15 working days of an occupationally exposed employee's request.

Responsible person/party: _____

Schedule of Implementation

Date Required	ECP Element	Date Completed
_____	_____	_____
_____	_____	_____
_____	_____	_____
_____	_____	_____
_____	_____	_____

Exposure Determination

Each company or organization having employees with Exposure must be identified and included in the Exposure Control Program. This process of identification is known as *exposure determination* and requires the creation of two lists:

1. The first list contains the job classifications in which *all* employees in a particular job classification have Exposure—for example, physicians and nurses, laboratory technicians, and possibly first-aid responders.
2. The second list contains the job classifications in which only *some* of the employees have Exposure, along with the associated tasks or procedures in that job classification in which Exposure may occur—for example, security personnel: Performing CPR.

Exposure Determination List

A.1. Indicate (with an X) those job classifications in which *all* employees may have occupational exposure to bloodborne pathogens:

[X] Physician	[X] First-aid responders
[X] Nurse	[X] Selected management
[X] Physician's assistant	[ ] _____
[X] Laboratory technician	[ ] _____

List tasks/procedures associated with occupational exposures:
First-aid responders: Clean up after accident or spill

_____
_____
_____

A.2. Complete evaluations for employees in A.1. (*Note*: All employees listed must be offered full protection under the OSHA standard.)

_____	_____	_____
_____	_____	_____
_____	_____	_____
_____	_____	_____
_____	_____	_____

B. List job classifications in which some (but not all) employees have occupational exposures to bloodborne pathogens (use additional pages as needed).

Job Classification/ Position	Tasks/Procedures Associated with Occupational Exposures
_____	_____
_____	_____
_____	_____
_____	_____
_____	_____
_____	_____
_____	_____
_____	_____
_____	_____
_____	_____
_____	_____
_____	_____

C. List employees who may have occupational exposures in the job classifications identified in Section B (use additional pages as needed). (All employees listed must be offered full protection under the OSHA standard.)

Name	Job Classification/Position	Date of Completion

ENGINEERING AND WORK PRACTICE CONTROLS

A number of preventive and protective actions must be taken to eliminate or minimize occupational exposures to bloodborne pathogens. This section addresses compliance methods known as Universal Precautions, Engineering Controls, and Work Practice Controls.

Universal Precautions

Universal Precautions is an OSHA-mandated system of infection-control techniques which operates under the premise that the only prudent way of protecting employees from bloodborne pathogens is to treat all body fluids (OSHA-defined) as potentially infectious. Attendant with this philosophy is the necessity to protect all potential portals of entry (e.g., mucous membranes of eyes, nose, mouth; skin) exposed to body fluids in the course of performing job functions (phlebotomy, wound care, PAP smears, etc.).

Engineering and Work Practice Controls

Engineering Controls and Work Practice Controls must be employed as a means of reducing occupational exposures. Engineering Controls accomplish this by using specifically designed equipment and facilities, while Work Practice Controls reduce exposure by altering the manner in which a task is performed.

Engineering controls	Work practice controls
Handwashing facilities	Handwashing
Sharps	Use of sharps
Sharps containers	No recapping of needles
Red bags/containers (appropriately labeled cabinets and storage containers)	No eating, drinking, applying cosmetics, or handling contacts in prohibited areas

Note: Each company or organization is responsible for assuring that Engineering Controls are made available, utilized, and maintained on a regular schedule; that employees follow the Work Practice Controls identified for the job or task; and that employees are trained in the use of Engineering Controls and safe work practices.

Handwashing and Handwashing Facilities

Readily accessible handwashing facilities must be provided for occupationally exposed employees. When this is not feasible, antiseptic towelettes or antiseptic hand cleansers used with cloth or paper towels followed by soap-and-water handwashing must be used (e.g., emergency response in remote area of plant). Employees who use gloves or other personal protective equipment must wash their hands immediately (or as soon as possible) after their removal. Similarly, skin must be washed with soap and water and mucus membranes flushed with water immediately after contact with blood or other potentially infectious agents. While not required, antimicrobial soaps (e.g., Dial, Betadine) are recommended.

Needles and Sharps

Careful handling and disposal of needles and other sharp implements or objects (e.g., scissors, broken glass) cannot be overemphasized. Such practices may represent the single most important protection against bloodborne pathogen transmission. Contaminated needles and sharps shall not be capped, bent, broken, cut, or manipulated (unless, in the case of recapping, specially designed devices for this purpose are used) and must be disposed of in sharps containers as soon as possible after use. Retrieval from sharps containers is prohibited. Containers must be (1) closeable, (2) puncture resistant, (3) leakproof, and (4) color-coded (red or labeled with biohazards symbol).

Food, Cosmetics, Contacts

Eating, drinking, smoking, applying cosmetics or lip balm, handling contact lenses, and the presence of food or beverages are prohibited in areas where Exposure can occur.

Procedures Involving Blood or Other Potentially Infectious Materials; Blood Specimen Handling and Transportation

All procedures must be performed in a manner that minimizes the likelihood of splashing, splattering, spraying, and forming droplets of blood or other potentially infectious materials (hereinafter "Blood"; e.g., the use of microhematocrit machine with a breakproof observation window). Mouth pipetting is forbidden!

Engineering Controls Checklist

List Engineering Controls (e.g., handwashing facilities; sharps containers; blood specimen containers; infectious waste containers; handling, storage, and transportation containers) in use in this facility. Complete or check all that apply.

[ ] Handwashing facilities
[ ] Sharps containers
[ ] Infectious waste containers
[ ] Handling, storage, and transportation containers
[ ] Regular schedule to examine and maintain or replace Engineering Controls, as needed
[ ] Handwashing facilities readily accessible to occupationally exposed employees (or, when unavailable, antiseptic towelettes or antiseptic hand cleansers with cloth or paper towels followed by soap-and-water handwashing)
[ ] All containers that may be used for storage, transport, or shipping of blood or other potentially infectious materials appropriately labeled or color-coded (biohazard warnings or red bags/containers), leakproof, and themselves not contaminated on the outside
[ ] Review of Engineering Controls at least [_____] time(s) per year

Other (list):

[ ] _____
[ ] _____
[ ] _____
[ ] _____
[ ] _____
[ ] _____

Inspection and Maintenance Schedule Form

- *Handwashing facilities*: Sinks with soap (preferably antimicrobial) are available at the following locations for use by occupationally exposed employees: Numerous locations throughout facility

_____
_____
_____
_____
_____
_____
_____
_____

Wherever handwashing facilities are not made accessible, antiseptic towelettes [X] and/or antiseptic hand cleansers and cloth or paper towels [__] are available for use at the following locations (e.g., in first-aid kits):

Antiseptic towelettes available in first-aid boxes located in the rear of the plant and in the office area.

_____

_____

_____

Inspection/maintenance schedule of handwashing facilities: ____[daily]____
Responsible person/party: _____[name]_____

- *Sharps containers* (leakproof, puncture-resistant, labeled):
  Locations:
  At first-aid station

_____

_____

_____

_____

Types of containers used:
See attached.

_____

_____

_____

[ ] Replaced/disposed of when three quarters full?
Responsible person/party: _____[name]_____

- *Other containers for contaminated materials,* such as glass, garbage pails, laundry (labeled, leakproof, can readily hold contents):
  Locations:
  First-aid station in office and breakroom

_____

_____

_____

Types of containers used:
Hazardous waste bags

_____

_____

_____

Method of disposal:

_____

Responsible person/party: _____[name]_____
[ ] Employees have been informed of and trained in the use of all of the above engineering controls.
Responsible person/party: _____[name]_____

Work Practice Control Checklist

Complete or check all that apply.

Handwashing

Employees are informed and trained that they must:

[ ] Wash their hands as soon as possible after removal of gloves and other personal protective equipment (e.g., goggles, face masks).

[ ] Wash their hands with soap and water or flush mucous membranes with water as soon as possible after occupational exposure.

Sharps

[ ] No contaminated needles are bent, recapped, or removed.

[ ] All contaminated, disposable, and reusable sharps (e.g., scissors) are immediately placed in puncture-resistant, labeled or color-coded, leakproof containers while awaiting disposal or reprocessing/decontamination.

[ ] Employees have been instructed not to overfill containers.

Personal Hygienic Practices

[ ] Eating, drinking, smoking, applying cosmetics or lip balm, and handling contacts are prohibited in areas where occupational exposure might occur.

[ ] Food and drink are not kept in refrigerators, freezers, shelves, cabinets, or countertops where potentially contaminated materials are or may be present.

Miscellaneous Work Practices

[ ] All procedures involving blood or other potentially infectious materials are performed in a manner that minimizes splashing, spraying, spattering, and generation of droplets.

[ ] Mouth pipetting is prohibited (if applicable).

[ ] Containers for storage, transport, or shipping of potentially infectious materials are labeled or color-coded and closed prior to leaving the facility (see Hazard Communication).

[ ] Containers with outside contamination are placed within other noncontaminated containers meeting the same required specifications.

[ ] Equipment that may have become contaminated with blood is inspected and decontaminated prior to servicing or shipping (e.g., flexible sigmoidoscope).

[ ] Work Practice Controls are reviewed on a [yearly] basis.

[ ] Employees have been informed and trained in all of the above Work Practice Controls in this checklist.

Responsible person/party: _____[name]_____

PERSONAL PROTECTIVE EQUIPMENT

Personal protective equipment (PPE) must be used when the potential for occupational exposure exists. EMS, Inc., will provide a full complement of PPE.

AVAILABILITY

Appropriate, properly sized PPE must be provided without cost and made readily accessible to employees when there is the potential for Exposures; for example, gloves, gowns, lab coats, eye protection, masks, faceshields, oropharyngeal airways, resuscitation bags, pocket masks, and overlay barriers are considered appropriate if they protect against infectious materials reaching mucous membranes, skin, and work or street clothes. Hypoallergenic gloves, glove liners, and other alternatives must be made readily accessible to employees who are allergic to standard gloves. It is the company or organization's responsibility to assure that the employee is trained on how to use, and the uses of, the appropriate PPE provided. For examples of recommended PPE by task, see attachment A, this section.

Gloves

Gloves (preferably latex) must be worn when it can be reasonably anticipated that an employee may have hand contact with blood, mucous membranes, and nonintact skin or may handle/touch contaminated surfaces or items. Good hygienic practice also dictates that employees who provide health care and have potentially infectious skin lesions or breaks in their skin protect themselves and their patients by using gloves.

Masks; Eye and Face Protection

Masks in combination with goggles or glasses with solid sideshields or chin-length faceshields must be worn whenever splashes or droplets of blood/infectious materials may be generated and eyes, nose, or mouth contamination might occur.

Protective Clothing

Gown, aprons, lab coats, and clinic jackets must be worn in occupational exposure situations according to the task and degree of exposure anticipated.

Respiratory Equipment

Respiratory equipment such as resuscitation bags, pocket masks with one-way valves, oropharyngeal airways, or overlay barriers must be used so that mouth-to-mouth resuscitation is avoided.

Cleaning, Laundering, Disposal

All personal protective equipment must be removed prior to leaving the work area and placed in the appropriate bags or containers for washing, decontamination, or disposal. Should garments be penetrated by blood or other potentially infectious materials, they must be removed immediately and placed in appropriate containers for contaminated laundry. PPE cleaning, repair, and replacement must be performed at no cost to the employee. Home laundering is forbidden. Disposable gloves shall be replaced as soon as possible when contaminated, torn, or punctured and cannot be washed for reuse (utility gloves in good condition may be decontaminated and reused).

Note: Training in the proper selection, use, and disposal/decontamination of PPE is of critical importance. Personnel responsible for such PPE training must be qualified and preferably should be experienced in their use. For example, the technique of applying and removing latex gloves to avoid self-contamination should be demonstrated and must be understood and followed by all who may use this form of PPE.

Replacement Personal Protective Equipment

Replacement PPE is available at first-aid stations or in the office.
Contact person: _____
Training responsible party: _____

Attachment A to Personal Protective Equipment

Recommended PPE by task

Task/activity	Disposable gloves	Gowns	Masks	Protective eye wear
Measuring blood pressure, temperature	No	No	No	No
Bleeding control (spurting blood)	Yes	Yes	Yes	Yes
Bleeding control (minimal bleeding)	Yes	No	No	No
Handling and cleaning contaminated instruments	Yes	No, unless soiling is likely	No	No

Availability

The following types of PPE (checked) are available in appropriate sizes for use by employees with potential for occupational exposure:

[ ] Gloves (utility)   [ ] Resuscitation bags
[ ] Gloves (latex and/or vinyl)   [ ] Pocket masks
[ ] Lab/clinic coats   [ ] Overlay barriers
[ ] Gowns   [ ] Other (list):
[ ] Eye protection (goggles or   _____
glasses with sideshields)   _____
[ ] Masks   _____
[ ] Faceshields   _____
[ ] PPE is provided without cost and made readily accessible

Gloves

[ ] Gloves are worn when it can be reasonably anticipated that an employee may have hand contact with blood or other potentially infectious materials, mucous membranes, and nonintact skin and during procedures requiring the handling or touching of contaminated surfaces or items.

[ ] Employees who use gloves are trained on when to use them, how to remove them without contaminating themselves, and how to appropriately discard and, in the case of utility gloves, decontaminate them.

Gloves are used during (check all that apply):

[ ]   First aid (bleeding)

[ ]   Decontamination (spills)

Other:

[ ]   _____

[ ]   _____

[ ]   Gloves are worn by employees who provide health care to others when the employees have potentially infectious skin lesions or breaks in the skin of their hands.

[ ]   Hypoallergenic gloves, glove liners, or other alternatives are made readily accessible to employees who are allergic to standard gloves.

[ ]   Disposable gloves are discarded into appropriate, nearby containers as soon as possible after contamination. Disposable gloves are never washed for reuse.

[ ]   Utility gloves are available for housekeeping/decontamination.

Gloves appropriate to the task may be obtained by

[ ]   Issuance by or on request of the manager/supervisor

[ ]   Use of first-aid/emergency kits that contain them

[ ]   Other: _____

If gloves are not used, explain reason why:

_____
_____
_____
_____

Facial and Respiratory Protection

[ ] Masks in combination with goggles or glasses with solid sideshields or chin-length faceshields are used whenever splashes or droplets of blood/ infectious materials might contaminate eyes, nose, mouth, or facial skin (e.g., control of spurting blood).

[ ] Resuscitation bags, pocket masks (with one-way valves), or overlay barriers are used to avoid mouth and respiratory system contamination during cardiopulmonary resuscitation (CPR).

[ ] Work practices specify under what circumstances, if any, facial protection may be necessary.

Facial/respiratory personal protective equipment may be obtained by
[ ] Issuance by or on request of the manager/supervisor
[ ] Use of first-aid/emergency kits which contain them
[ ] Other: _____
If facial/respiratory protection is not used, explain reason why:

_____
_____
_____
_____

Protective Clothing

[ ] Gowns, aprons, and lab coats are available in occupational exposure situations according to task and degree of exposure anticipated (for example, spurting blood, sigmoidoscopic evaluations, decontamination procedures involving large spills of blood).
Protective clothing may be obtained by
[ ] Issuance by or on request of the manager/supervisor
[ ] Other: _____
[ ] Contaminated protective clothing is placed in appropriate containers as soon as possible after contamination and discarded, decontaminated, or laundered.
Laundering of soiled protective clothing is performed by
_____[company/city/state]_____

(*Note*: Employees are not permitted to launder their own contaminated clothing.)

If protective clothing is not utilized, explain reason why:

_____
_____
_____
_____

[ ] Employees have been informed and trained in the selection, use, and disposal/decontamination of the above personal protective equipment where applicable.
Responsible person(s)/party(s): _____[name]_____
Training by: _____[name]_____
The availability and use of personal protective equipment are reviewed on a
_____[yearly]_____ basis by _____[name]_____.

HOUSEKEEPING

Note: Many locations use outside contractors for their security, housekeeping, laundry, and other services, the performance of which may have the potential for occupational exposures. Therefore, it is advisable to obtain a written agreement that clarifies which organization will be responsible for performing particular tasks having the potential for occupational exposure and under what circumstances and under whose auspices such tasks will be performed. In most instances, it is the company's responsibility to inform outside contractors or their employees about the potential for exposure to contaminated materials (e.g., janitorial staff) and it is the outside contractor's responsibility to provide the full protection of the Bloodborne Pathogens standard, including, but not limited to, education, training, and personal protective equipment for their own employees.

General

All worksites must be kept in a clean and sanitary condition. A written schedule for cleaning and a procedure for decontamination of worksites must be available and appropriate for each:

- Location (breakroom, main plant)
- Type of surface to be cleaned (counters, floor tile, or carpet)
- Type of soiling present (bloody emesis, small splatter of blood)
- Tasks/procedures performed

Equipment, Environmental, and Working Surfaces

All equipment, environmental, and working surfaces must be cleaned and decontaminated with an appropriate disinfectant (e.g., a 1:10 bleach-to-water solution; virucidal or tuberculocidal solutions) as soon as possible after contact with "blood or other potentially infectious materials" ("Blood"), upon completion of procedures that could give rise to contaminated surfaces, and at the end of the work shift if surfaces may have been contaminated since the last cleaning.

Note: Disinfectants should be selected based on the Centers for Disease Control Guidelines, MMWR1989, 38 (No. S-6), pages 1–37. Examples include one part household bleach to ten parts water, Steriphene II disinfectant (may be safely used on carpets), Cidex (for cleaning sigmoidoscopic equipment).

Similarly, protective coverings (e.g., plastic wrap, aluminum foil) must be appropriately discarded as soon as possible after visible contamination or at the end of the work shift if contamination could have occurred. All reusable receptacles (bins, pails, pans) must be inspected, decontaminated, and cleaned on a regularly scheduled basis if there is a reasonable likelihood of their being contaminated by Blood and as soon as possible after visible contamination.

Glassware/Reusable Sharps

Broken glassware that may be contaminated must not be picked up directly by hand. A mechanical means of picking up such material (tongs, brush and dust pan, or forceps) must be used in such cases. The mechanical devices themselves must then be decontaminated. Reusable sharps must not be stored or decontaminated in a manner that requires employees to reach by hand into the containers in which they have been placed.

Regulated Waste

Contaminated sharps: Containers for sharps shall be discarded as soon as possible in containers that are (1) closeable, (2) puncture resistant, (3) leakproof (sides and bottom), and (4) properly labeled or color-coded.

Containers for contaminated sharps shall be (1) easily accessible, (2) located as close as possible to the area where they are used or where sharps might be found (e.g., laundries), and (3) kept upright.

When moving sharps containers, they must be (1) closed just prior to moving, and (2) placed in secondary leakproof/puncture-resistant containers (if leakage or puncture is possible), which must also meet all of the required characteristics of the primary container.

Reusable containers must not be opened, emptied, or cleaned in a manner that poses a risk of occupational exposure or percutaneous (through the skin) injury.

Other contaminated waste: Other contaminated waste must be handled, contained, and disposed of in the same meticulous fashion as for "contaminated sharps," except for the lack of need for puncture-resistant containers. Routine checks must be performed as necessary for outside contamination and secondary containers used to enclose outside-contaminated primary containers (see Engineering and Work Practice Controls).

Note: Disposal of regulated waste must be in accordance with applicable federal, state, and local regulations.

Laundry: Contaminated laundry must be (1) handled as little as possible, (2) bagged at the location where it is used (sorting is not permitted), and (3) placed and transported in properly labeled or color-coded containers which prevent soak-through or leakage. Employees who have contact with contaminated laundry must wear protective gloves and other appropriate personal protective equipment (e.g., gowns or aprons). Companies or organizations must use bags meeting the requirements for "other contaminated waste," above, prior to offsite shipping of contaminated laundry.

Disposal

When necessary, EMS, Inc., will contact BFI for disposal of contaminated waste.

Attachment A to Housekeeping

Contaminated spills: Recommended procedures
[ ] Secure area.
[ ] Contact [name] _____.
[ ] Determine decontamination equipment necessary:
Personal protective equipment (latex gloves [ ], masks [ ], goggles [ ], gowns [ ],
  booties [ ]; other _____ [ ]).

Absorbent materials (e.g., paper towels, commercial absorbent agents) (list):

[ ] ——————————————— [ ] ———————————————
[ ] ——————————————— [ ] ———————————————
[ ] ——————————————— [ ] ———————————————
[ ] ——————————————— [ ] ———————————————

[ ] 1:10 bleach solution

Other disinfectant solutions (EPA-defined, tuberculocidal) (list):

[ ] ——————————————— [ ] ———————————————
[ ] ——————————————— [ ] ———————————————

[ ] Appropriately labeled, puncture-proof/leakproof containers to receive contaminated materials

[ ] Appropriate containers for processing or decontaminating housekeeping/ personal protective equipment used

Decontamination procedure

[ ] Put on appropriate personal protective equipment.

[ ] Cover spill with absorbent material(s).

[ ] Absorb and containerize grossly contaminated materials.

Note: If glass or sharps are present, do not remove directly using hands; use a dust pan and brush and place in an appropriately labeled, puncture-resistant container.

[ ] Disinfect remaining contaminated materials with appropriate chemical agent following manufacturer's instructions.

[ ] Wash/mop area with clean water and/or other appropriate cleansing agent.

[ ] Discard disposable PPE and equipment; containerize all reusable PPE and equipment for appropriate decontamination.

[ ] Wash hands thoroughly.

Housekeeping checklist

Schedule:

[ ] A written schedule for cleaning and decontaminating work sites is available.

Housekeeping procedures:

– Cleaning

[ ] All equipment, environmental and working surfaces are cleaned using appropriate disinfectants as soon as possible after known contamination; at end of procedures or work shift if potential for contamination.

[ ] Reusable receptacles are inspected, decontaminated, and cleaned on a regularly scheduled basis and as soon as possible after visible contamination.

[ ] Protective coverings are discarded at the ends of work shifts in which contamination could have occurred and as soon as possible after visible contamination.

– Broken glass

[ ] Broken glass is never picked up directly by hand.

[ ] Mechanical aids (tongs, brush and dust pan, or forceps) are used and appropriately discarded or disinfected before reuse.

− Sharps containers, reusable sharps containers; other contaminated waste

[ ] Reusable sharps containers are never entered by hand.

[ ] Contaminated sharps containers are (1) closeable, puncture resistant, and leakproof (sides and bottom); and (2) properly labeled or color-coded, easily accessible, located as close as possible to the area where they are used or where sharps may be found, and kept upright.

[ ] Sharps containers are not overfilled and are routinely replaced when no more than three quarters full.

[ ] Sharps containers are closed prior to moving.

[ ] Reusable containers are not opened, emptied, or decontaminated in a risky manner.

[ ] Other, non-sharps, contaminated waste is meticulously handled, contained, and disposed of in accordance with applicable federal, state, and local regulations.

− Laundry

Contaminated laundry is

    [ ] Handled as little as possible.

    [ ] Bagged at the location where it is used (sorting is not permitted).

    [ ] Placed and transported in properly labeled or color-coded containers that prevent soak-through or leakage.

    [ ] Gloves and, when necessary, gowns or aprons are used when handling contaminated laundry.

    [ ] Contaminated laundry is appropriately bagged and labeled (biohazard sign) prior to offsite shipping.

Miscellaneous

− Written/verbal understanding: There is a written or verbal understanding of the potential occupational exposure hazards, decontamination procedures, and responsibilities of the company and its contracted services:

Laundry service (name):

[ ] ——————————————————————————————

Other services (name):

[ ] ——————————————————————————————

[ ] ——————————————————————————————

[ ] ——————————————————————————————

Person/party responsible for documenting that the above-mentioned written or verbal understanding has been provided or discussed with contracted services (provide name of person):

[ ] ——————————————————————————————

[ ] Employees have been informed of, and trained in complying with, all of the above housekeeping requirements.

Person(s)/party(s) responsible:

_____

_____

_____

Housekeeping schedule
Worksite(s) (location in facility):
  Complete facility cleaned and disinfected daily in accordance with regulations

_____

Type of surface:
  Varied surfaces

_____

Methods of decontamination and type of disinfectant (use only approved disinfectants):
  See attached list of USDA-approved disinfectants.

_____

Frequency of cleaning/decontamination:
  Daily

_____

Responsible person(s)/party:

_____

Manager

HEPATITIS B VACCINE AND VACCINATION SERIES AND POSTEXPOSURE EVALUATION

Note: Successful compliance with this section must include the completion of all forms. The OSHA Hepatitis B Vaccine Declination Form is mandatory.

Hepatitis B Vaccination

The hepatitis B vaccine is a safe, effective immunization against the hepatitis B virus and must be offered free of charge to all employees who may have Exposure after they have participated in a required training program. Employees with Exposure have the option of declining the vaccine; however, any at-risk employee who wishes not to receive the vaccine must sign a Hepatitis B Vaccine Declination Form. If the employer later decides to receive the vaccine, it must still be offered free of charge. The initial vaccine offering must be made no later than July 6, 1992. New employees or employees whose job tasks have changed such that they are now occupationally exposed must be offered the vaccination within 10 working days of beginning the new job task.

Hepatitis B vaccination must be made available to eligible employees at a reasonable time and place and must be given or supervised by a licensed healthcare professional (physician, nurse). EMS, Inc., has designated Dr. Doe at the Family Clinic as its designated physician. The vaccination series currently consists of three separate injections given over a six-month time period. More than 90 percent of those vaccinated according to the recommended dosage schedule will develop immunity to the hepatitis B virus. *Note*: The healthcare professional administering the vaccination series must be provided a copy of the OSHA standard.

Employees who have received the HBV vaccination series, are immune to HBV, or have a medical contradiction to receiving the vaccination series are exempt from the requirements of this section. At the present time, booster doses of vaccine are not recommended. If, in the future, a booster dose(s) of hepatitis B vaccine is recommended by the U.S. Public Health Service, such booster dose(s) must be made available free of charge to employees at risk.

Postexposure Evaluation and Follow-Up

Any employee who has had an "exposure incident" must be offered a postexposure evaluation and follow-up (including any indicated laboratory tests), free of charge, at a reasonable time and place, under the supervision of a licensed healthcare professional (physician, nurse) and according to current U.S. Public Health Service recommendations. Following an exposure incident, the confidential medical evaluation and follow-up must be made immediately available to the affected employee. Such evaluation and follow-up must include all of the following:

- Documentation of route(s) of exposure and circumstances under which the incident occurred.
- Identification and documentation of the source individual where feasible (unless prohibited by law).
- Immediate testing of the source individual's blood to determine HBV and HIV status, unless already known (after obtaining written consent; if consent is not obtainable, this must be documented and placed in the employee's medical record). Results of testing must be made available to the affected employee, along with applicable laws/regulations prohibiting disclosure of identity and clinical status of the source individual.
- Determination of affected employee's HBV and HIV status. Blood testing must be initiated as soon as possible after consent is obtained. The employee may elect to have baseline blood drawn but can delay a decision concerning testing on that blood specimen. In such an instance, the blood specimen must be preserved for at least 90 days, whereupon it may be discarded.
- Postexposure prophylaxis, if indicated, per U.S. Public Health Service guidelines.
- Employee counseling.
- Medical evaluation of any employee-reported illnesses.
- Inclusion of exposure incident on OSHA 200 Injury/Illness Log.

Information Provided to the Healthcare Professional

The company or organization is responsible for providing the healthcare professional who will be performing the employee's hepatitis B vaccination with a copy of the OSHA Bloodborne Pathogens standard. In addition, the healthcare professional evaluating an employee after an exposure incident must be provided:

- Copy of the Bloodborne Pathogens standard
- Copy of this written program

- Description of the employee's duties as they relate to the exposure incident
- Documentation of the routes of exposure and circumstances under which exposure occurred
- Results of the source individual's blood testing, if available
- All medical records relevant to the appropriate treatment of the employee, including vaccination status

Healthcare Professional's Written Opinion

The company or organization must obtain and make available to the exposed employee a copy of the evaluating healthcare professional's written opinion within 15 days of the completion of the evaluation. The written opinion shall be limited to indicating if the employee should receive hepatitis B vaccination and if such vaccination has been given. The healthcare professional's written opinion concerning postexposure evaluation and follow-up shall be limited to an indication that the exposed employee has been informed of the results of the evaluation and has been told about any medical conditions resulting from Exposure that may require further evaluation or treatment. All of the findings or diagnoses shall remain confidential and not be included in the written report. EMS, Inc., will maintain all medical reports in the individual employee's medical file.

Postexposure Response Guidelines

Decontamination:
- Skin: Wash thoroughly with soap and water.
- Mucous membranes (eyes, nose, mouth): Rinse thoroughly with water or normal saline.
- Environmental surface: Wash with solution of 1:10–1:100 household bleach and water or other appropriate disinfectant.

Reporting of exposure incident to designated individual:
- Document exposure, including route (i.e., needlestick or absorption through mucous membranes [eyes, nose, mouth]).
- Complete OSHA 200 Injury/Illness Log.

Healthcare professional evaluation:
- Employee medical evaluation
- Review of exposed employee's medical record relevant to exposure incident (i.e., hepatitis B vaccine status)
- Identification of source individual and, if possible, any HIV and HBV testing, once consent is obtained
- Employee blood testing for HBV and HIV, once consent is obtained (both HBV and HIV prophylaxis needs to be considered in the medical evaluation of any employee who has occupational exposure)

HIV postexposure prophylaxis:
Consider zidovudine (AZT).

HBV postexposure prophylaxis:

Source status	Tested positive	Tested negative	Not tested
Unvaccinated	Administer HBIGx1[a] and initiate hepatitis B vaccine series	Initiate hepatitis B vaccine series	Initiate hepatitis B vaccine series
Previously vaccinated, known responder	Test exposed person for anti-HBsAG; if adequate,[b] no treatment; if inadequate, give hepatitis B vaccine booster	No treatment	No treatment
Known non-responder	Repeat HBIGx2 or HBIGx1 hepatitis B vaccine	No treatment	If known high-risk source, may treat as if source were HBsAG positive
Response unknown	Test exposed person for anti-HBsAG; if inadequate, give HBIGx1, plus hepatitis B vaccine series; if adequate, no treatment	No treatment	Test exposed person for anti-HBsAG; if inadequate, give hepatitis B vaccine booster dose; if adequate, no treatment

[a] Hepatitis B immune globulin (HBIG) dose 0.06 mL/kg intramuscularly.
[b] Adequate anti-HBsAG (hepatitis B surface antigen) is greater than or equal to 10 milli-international units.

Medical counseling and follow-up:
   Written opinion of Dr. Doe will be sent to EMS, Inc., which will forward it to the employee within 15 days of completion of evaluation.

Employees Eligible for, and Offered, Hepatitis B Vaccine and Vaccination Series

Employee Name	Date Offered	Accepted? (Y/N)	If Declined, Is Declination Form Signed? (Y/N)
_____	_____	____	____
_____	_____	____	____
_____	_____	____	____
_____	_____	____	____
_____	_____	____	____
_____	_____	____	____
_____	_____	____	____
_____	_____	____	____
_____	_____	____	____
_____	_____	____	____
_____	_____	____	____
_____	_____	____	____

Hepatitis B Vaccine Declination Form ECP VIII-2 (Mandatory)

I understand that due to my occupational exposure to blood or other potentially infectious materials I may be at risk of acquiring hepatitis B virus (HBV) infection. I have been given the opportunity to be vaccinated with the hepatitis B vaccine, at no charge to myself; however, I decline hepatitis B vaccination at this time. I understand that by declining this vaccine, I continue to be at risk of acquiring hepatitis B, a serious disease. If, in the future, I continue to have occupational exposure to blood or other potentially infectious materials and I want to be vaccinated with the hepatitis B vaccine, I can receive the vaccination series at no charge to me.

Name (please print): _____

Title: _____

Signature: _____

Date: _____

Site/facility location: _____

Place in confidential medical record.

Employee Exposure Incident Checklist

Note: This checklist is to be used when an employee sustains an "exposure incident" (see Definitions section, above).

Employee actions:
   [ ] Decontamination of skin, mucous membranes, or environmental surface
   [ ] Notification of exposure incident to designated individual
Responsible person(s)/party(s) actions:
   [ ] Documentation of exposure incident by completing Employee Exposure Incident Record
   [ ] Completion of OSHA 200 Injury/Illness Log
   [ ] Identification and documentation of source individual where feasible, unless prohibited by law
   [ ] HBV and HIV testing of source individual's blood after obtaining written consent (unless already known); document if source individual's consent is not obtained
   [ ] Results of source individual's test results made available to affected employee with prohibitions against disclosure of source individual's identity or clinical status
   [ ] Determination of affected employee's HBV and HIV status (complete Employee Exposure Incident Record)
Information provided to the health professional evaluating an affected employee after exposure incident:
   [ ] Copy of Bloodborne Pathogens standard
   [ ] Employee duties as they relate to exposure incident
   [ ] Routes of exposure and circumstances under which exposure occurred

[ ] Results of source individual's blood testing, if available
[ ] All relevant medical records
[ ] Medical evaluation and treatment as outlined in the postexposure response guidelines section, above.
[ ] Written medical opinion sent to company or organization management and forwarded to affected employee within 15 days of completion

Employee Exposure Incident Record

Date and time of exposure incident:

_____

Name of exposed employee:

_____

Exposure incident reported to:

_____

Route(s) of exposure:

_____

Circumstances of exposure incident (use additional sheet, if necessary):

_____
_____
_____

Medical evaluation performed by:

_____

Employee vaccination status:

_____

Employee HBV/HIV status:

_____

Medical evaluation and treatment (including postexposure prophylaxis and counseling):

_____

Written opinion of medical evaluation sent to company or organization management and forwarded to exposed employee on (date):

_____

Place in confidential medical record.

Hazard Communications

Warning labels and signs must be used to identify items that can pose a hazard. These labels and signs will ensure that anyone who may come in contact with objects so identified know that they must handle the objects with care. OSHA has chosen the fluorescent biohazard sign as the appropriate symbol and red as the designated hazard color code. Labels may be attached by string, wire, adhesive, or any other method that will not allow them to fall off accidentally. They can also be a part of a container itself. Red bags or red containers may be substituted for labels.

Biohazard Sign

Containers of regulated waste, refrigerators, or freezers that hold potentially infectious materials (e.g., blood), and other containers used to transport or store blood or infectious materials must be labeled with the biohazard sign or be color-coded (red bags or red containers). Contaminated equipment must be labeled so as to designate which portions, if any, remain contaminated. Individual containers that are placed inside another labeled container for storage or transport do not need to be labeled separately. Regulated waste that has been decontaminated does not need any labeling or color-coding.

Hazard Communications Checklist

[ ] Warning labels are affixed to containers of regulated biomedical waste, refrigerators, and freezers containing blood or other potentially infectious material. Containers used to store, transport, or ship blood or other potentially infectious materials are labeled in a similar fashion.

[ ] Warning labels use the OSHA-designated biohazard sign that is fluorescent orange or orange-red with letters or symbols in a contrasting color.

[ ] Warning labels are either a part of the container itself or attached by string, wire, adhesive, or any other method that will not allow them to fall off accidentally.

[ ] Red bags or red containers are substituted, where appropriate, for labels. The hazards of blood and other potentially infectious material have been communicated to all employees who have occupational exposure in this facility and the use of appropriate warning labels and signs has been explained.

Responsible person(s)/party: _____[name]_____

TRAINING AND EDUCATION

The standard has mandated that EMS, Inc., provide all their employees determined to have occupational exposure an educational training program, free of charge and conducted during normally scheduled working hours. It may be helpful when conducting these training sessions to prepare examples of signs and labels that are required (e.g., biohazard labels), have available articles of personal protective equipment to demonstrate their proper use (e.g., gloves, gowns, face masks, mouthpieces for CPR), and demonstrate the work practices that are essential to this program (e.g., disposal of sharps into sharps container). Training videos are available. These may be used as an aid in the training session but are not to be used in place of an individual trainer. The training program must be provided to an employee at the time of initial employment or before June 4, 1999, and must be provided at least annually thereafter to all occupationally exposed employees. Additional training must be provided when new or modified tasks or procedures may affect an employee's occupational exposure. These training records must be

• Kept for three (3) years from the date of the training session.
• Kept available upon request to all employees or their representatives.
• Made available upon request to OSHA.

Examples of training program videos include:

Bloodborne Pathogens	Preventing Bloodborne Disease
Summit Training Service, Inc.	Tel-A-Train
(616)784-4500; (800)842-0466	(615)266-0113
Time: 12 minutes	Time: 18 minutes
Excellent for general workplace settings	Excellent, especially for first responders

---

EMS, Inc., Training Program Requirements

The training program provides the employee with the following information:

1. Accessible copy of the regulatory text of this standard and an explanation of its contents
2. General explanation of the epidemiology and symptoms of bloodborne diseases
3. Explanation of the modes of transmission of bloodborne pathogens
4. Explanation of the Exposure Control Plan and where the employee can obtain a copy of the written plan
5. Explanation of recognizing tasks and other activities that may involve exposure to potentially infectious materials
6. Explanation of the use and limitations of methods to prevent or reduce exposure, including appropriate Engineering Controls, Work Practice Controls, and personal protective equipment
7. Information on the types, proper use, location, removal, handling, decontamination, and disposal of personal protective equipment
8. Explanation of the basis for selection of personal protective equipment
9. Information on the hepatitis B vaccine, including information on its efficacy, safety, and method of administration; benefits of being vaccinated; and the vaccine and vaccination being offered free of charge
10. Information on the appropriate actions to take and persons to contact in an emergency involving blood or other potentially infectious materials
11. Explanation of the procedure to follow if an exposure incident occurs, including the method of reporting the incident and the medical follow-up that will be made available
12. Information on the postexposure evaluation and follow-up that the employer is required to provide for the employee after an exposure incident
13. Explanation of the signs and labels or color-coding required to identify potentially infectious materials
14. Opportunity for interactive questions and answers with the person conducting the training session

Bloodborne Pathogens Training Program
Bloodborne Pathogens Training Record

Facility location: Date:

Trainer's name:
Trainer's qualifications:

Content (check all that apply):
[ ] Items 1–14 listed in training program requirements section, above
[ ] Q and A with trainer
[ ] Post-training test evaluation

Instructional materials (check all that apply):
[ ] Training program document
[ ] Bloodborne Pathogens standard
[ ] Videocassette
[ ] Other:

Trainees (use additional pages, as needed):

Name	Job Classification/Title
See attached.	

Bloodborne Pathogens Training Program

A. Bloodborne Pathogens standard

   Copies of the *Bloodborne Pathogens Manual*, this training program, and the OSHA standard upon which they are based are located in the office and are readily available for use by all employees. Contact [name].

B. Epidemiology and symptoms of bloodborne diseases

   Disease-producing organisms (pathogens) may be found in the blood and certain other body fluids. The pathogens can be transmitted by various means, such as breaks in the skin caused by a needlestick or any sharp instrument, by absorption through skin that has been abraded, or absorption through mucous membranes of the eyes, nose, or mouth. Two disease-producing organisms that are of most concern are the hepatitis B virus (HBV) and the human immunodeficiency virus (HIV). Although the focus of OSHA's standard is directed toward HBV and HIV, it is important that you realize there are other potentially infectious organisms present in blood; following the instructions in this training program for protection against HBV and HIV will also provide you with risk reduction against all bloodborne pathogens.

   • Hepatitis B

     The acute and chronic consequences of hepatitis B virus (HBV) infection are major health problems in the United States. An estimated 200,000–300,000 new infections occurred annually during the period 1980–1991. Approximately 8700 healthcare workers each year contract HBV, and about 200 will die as a result. Transmission of HBV is by direct injection through the skin (e.g., needlestick injuries in healthcare workers; the use of shared needles and syringes by intravenous drug users), blood contamination of mucous membranes (e.g., eyes, mouth, nose), exchange of sexual fluids during intercourse (e.g., semen and vaginal secretions), and transmission from an infected mother to her newborn infant. Infection with HBV may result in a variety of clinical symptoms from asymptomatic infection to a mild, flu-like illness to a very severe infection that may result in debilitating hepatitis, cirrhosis of the liver, or primary liver cancer. An individual who becomes infected with HBV may develop a chronic form of hepatitis from which they never completely recover. Currently, an estimated 1.25 million persons in the U.S. have chronic HBV infection. Persons infected with HBV are capable of transmitting this organism during any of the clinical states already mentioned. Once an individual is infected with HBV, the presence of certain "markers" can be determined (HBsAG, hepatitis B surface antigen; HBeAG, hepatitis B envelope antigen; HBsAb, hepatitis B surface antibody). HBsAG and HBeAG may give us information that an individual is still in the infectious stage of disease and capable of transmitting the hepatitis virus. HBsAb may show that the infected individual has produced antibodies against the hepatitis virus, which indicates recovery from infection. These antibodies are "protective antibodies" and prevent the individual from becoming ill if exposed to HBV again. These protective antibodies are what is produced in individuals who receive the hepatitis B vaccination series. The risk of death and impairment of health resulting from acute and chronic hepatitis B infection is significant. Therefore, knowledge of the disease and prevention against it are important. The best defense against HBV is to receive the hepatitis B vaccination series.

- Human immunodeficiency virus

   The organism that causes the Acquired Immune Deficiency syndrome (AIDS) is the human immunodeficiency virus (HIV). This virus was isolated during 1983–1984. HIV transmission can occur by the same methods as described for HBV transmission. Symptoms of HIV infection vary according to disease progression. After infection with HIV, the individual may experience an illness that is characterized by fever, enlarged lymph nodes, muscle and joint pains, diarrhea, fatigue, and possibly a rash. This acute illness is usually self-limiting and resolves within two to four weeks. The majority of individuals with HIV develop antibodies that are antigenic markers that can be found in the blood after infection with HIV. These are not "protective antibodies" as previously discussed for hepatitis B virus infection. Most persons who are infected with HIV may remain without symptoms for months to years following infection; however, the majority will eventually develop AIDS. HIV infection affects the immune system such that the infected individual becomes susceptible to a wide range of illnesses. These may include infections from bacterial, fungal, and parasitic organisms that an individual with a normal immune system would only rarely experience. AIDS patients may also be susceptible to developing various cancers and quite often experience neurologic complications. Any of these clinical disorders can be aggressive, rapidly progressive, difficult to treat, and less responsive to traditional methods of therapy. Transmission of HIV from an infected individual may occur throughout any stage of the disease. There is no vaccine against the HIV; therefore, the best protection is close adherence to the specific measures outlined in this training program to reduce the risk of infection from all bloodborne pathogens.

C. Modes of transmission of bloodborne pathogens

   Bloodborne infections may enter the body in the work environment through a break in the skin caused by a needlestick or any sharp instrument that has been contaminated with potentially infectious materials; by absorption through skin that is inflamed, abraded, or cut; or by absorption through mucous membranes in the mouth, nose, or eyes. Nonoccupational transmission of bloodborne infections can also occur through high-risk behavior, such as intravenous drug abuse and certain sexual activities.

D. Exposure Control Plan

The Exposure Control Plan has been designed to protect healthcare workers against potentially infectious materials. The objective of the plan is to provide employees with the education and training necessary to reduce or eliminate exposure to bloodborne pathogens. The Exposure Control Plan consists of several parts, as follows.

- Exposure determination

    Certain employees, by virtue of their job classification or title, have occupational exposure to blood and other potentially infectious materials. These include physicians, nurses, laboratory technicians, or any other employees with a similar risk to exposure. Certain employees may not normally be exposed to potentially infectious materials, but as part of their job duties may occasionally have contact with blood or other potentially infectious materials. These include CPR-trained employees (e.g., supervisors) who, as part of their job duties, act as first-responders to emergencies in the work environment; housekeeping staff who clean medical clinics or other areas where occupational exposure may occur; or any other employee who may occasionally have contact with potentially infectious materials.

- Methods of compliance

    Universal Precautions must be observed to prevent contact with blood or other potentially infectious materials. Under circumstances in which differentiation between body fluid types is difficult or impossible, all body fluids shall be considered potentially infectious materials.

- Engineering Controls and Work Practice Controls

    Engineering and Work Practice Controls are used to eliminate or minimize employee exposure. Where occupational exposure remains after institution of these controls, personal protective equipment should also be used.

- Personal protective equipment (PPE)

    When there is potential for occupational exposure, the company will provide, at no cost to the employee, appropriate personal protective equipment such as, but not limited to, gloves, gowns, laboratory coats, faceshields or masks and eye protection, mouthpieces, resuscitation bags, pocket masks, or other ventilation devices.

- Housekeeping

    All worksites are to be maintained in a clean and sanitary condition. Each worksite will be cleaned on a specified schedule. General procedures for cleaning and disinfecting minor and large spills are outlined in Section J.

- Medical waste disposal

    The procedures for disposal of infectious waste will be designed to protect the health and safety of the healthcare employee while adhering to federal, state, and local regulations.

- Hepatitis B vaccination and postexposure evaluation and follow-up
    The Hepatitis B vaccination series will be offered free of charge to those employees who have been determined to have occupational exposure to blood and other potentially infectious materials. This vaccine series, along with the safe work practices covered in this training program, is one of the most effective ways to prevent transmission of HBV. In addition, a plan is in place should you have an exposure incident. This plan includes a medical evaluation, specific treatment when necessary, and medical follow-up.
E. Tasks/activities with occupational exposure
    The tasks and activities that may involve exposure to blood or other potentially infectious materials in a given job classification or job title are used to make the exposure determination of those employees at potential risk.
F. Methods of compliance
    Methods of compliance are various practices and/or procedures that are to be followed to minimize the potential for occupational exposure. Methods of compliance include: (1) Universal Precautions; (2) Engineering and Work Practice Controls; (3) personal protective equipment; (4) housekeeping; and (5) laundry.
    - Universal Precautions
        Universal Precautions is a system of infection control techniques developed by the Centers for Disease Control (CDC) designed to protect employees from exposures to blood and other potentially infectious materials. Because bloodborne infections can be transmitted from apparently healthy persons who have no outward signs or symptoms of a disease, you should therefore assume that *body fluids from all patients are potentially infectious.*
    - Engineering Controls and Work Practice Controls
        Engineering Controls are methods used to eliminate or minimize occupational exposure. These can include self-sheathing needles, conveniently placed sharps containers and red bag containers for disposal of potentially infectious materials, handwashing facilities, and appropriately labeled (e.g., biohazard warning) cabinets or storage containers for blood or other potentially infectious materials. Eating, drinking, smoking, applying cosmetics or lip balm, and handling contact lenses are prohibited in work areas where there is the likelihood of occupational exposure. Food and drink must not be kept in refrigerators, freezers, shelves, cabinets, or on countertops or benchtops where blood or other potentially infectious materials are present. Work Practice Controls reduce the likelihood of occupational exposure by assuring that procedures are properly performed:
        Handwashing facilities: Employees must wash their hands immediately, or as soon as possible, after removal of gloves or other PPE. Gloves should not be considered as a substitute for handwashing. If handwashing facilities are not readily available (e.g., when responding

to the scene of an emergency), an antiseptic hand cleanser with a cleaning cloth or paper towels or antiseptic towelettes will be provided in the emergency equipment and must be used. When antiseptic cleansers or towelettes are used, hands must be washed with soap and water as soon as possible.

Management of sharps: Sharps containers must be puncture-resistant, labeled with the universal biohazards symbol or color-coded in red, leakproof on the sides and bottoms, and translucent, to determine the amount of material in the container. Contaminated needles and sharps are not to be recapped or removed unless no alternative is available. If recapping or needle removal is to be done, it must be done with the use of a mechanical device or a one-handed technique.

Personal protective equipment (PPE): Where there is potential for occupational exposure, the company will provide at no cost to the employee appropriate PPE such as, but not limited to, gloves, gowns, laboratory coats, faceshields or masks and eye protection, mouthpieces, resuscitation bags, pocket masks, or other ventilation devices. When PPE use is required, it will be provided in appropriate sizes and will be readily available. For special needs (e.g., requiring the use of hypoallergenic gloves or powderless gloves), accommodations will be made and the equipment will be available.

EMS, Inc., will clean, launder, and dispose of PPE provided to employees, free of charge. *Employees are not allowed to launder their own PPE.* Never leave a work area without first removing any PPE and disposing of it in the appropriate containers.

Disposable gloves must be worn when it can be reasonably anticipated that you may have skin or mucous membrane contact with blood or other potentially infectious materials (e.g., when drawing blood, handling vaginal speculums, cleaning sigmoidoscope). As soon as a task is completed, gloves are to be removed and discarded in the appropriate containers, and hands must be washed. Disposable (single-use) gloves are not to be washed or decontaminated for reuse. For general housekeeping activities, utility gloves may be decontaminated for reuse; however, utility gloves must be discarded if they are cracked, peeled, torn, punctured, or in a state of deterioration.

Masks in combination with shields such as goggles or glasses with solid side panels or chin-length faceshields are worn whenever splashes, sprays, spatters or droplets of blood, or other potentially infectious materials may be anticipated and eye, nose, or mouth contamination could occur. Prescription eyeglasses with solid side-shields are appropriate.

Appropriate protective clothing may be gowns, aprons, lab coats, clinic jackets, or similar clothing. The choice of what clothing to wear will depend upon the task and risk of exposure anticipated (e.g., gloves only for blood drawing).

Mechanical respiratory assist devices such as mouthpieces, bag valve masks, pocket masks, and other such CPR aids will be available to all personnel who have trained in CPR for resuscitation use.

Housekeeping: All worksites are to be maintained in a clean and sanitary condition according to a schedule based upon location, type of surface, type of contamination, and task or procedures being performed. All equipment and surfaces must be cleaned and decontaminated after contact with blood or other potentially infectious materials. The use of disinfectants is based upon the Centers for Disease Control (CDC) guidelines and includes common household bleach and Cidex. Protective coverings such as plastic wrap, aluminum foil, or other imperviously backed absorbent paper may be used to cover equipment or environmental surfaces. Coverings are to be removed and replaced as soon as feasible (e.g., between patients) when they have become contaminated or at the end of the work shift, if they have become contaminated during the shift. PPE must be worn for all decontamination procedures. Employees are strictly prohibited from placing hands into receptacles to retrieve any materials. Broken glass or other sharp items should be removed using mechanical devices, such as a brush and dustpan or forceps. Mechanical devices are either disposed of or decontaminated after use. All receptacles (reusable) that have a reasonable likelihood for becoming contaminated with blood or other potentially infectious materials are inspected on a regular basis and decontaminated when appropriate.

G. Personal protective equipment

See personal protective equipment section, above.

H. Basis for selection of personal protective equipment

1. *Gloves*: In performing tasks where you can reasonably anticipate hand contact with blood or other potentially infectious materials, you must wear gloves.

2. *Face protection*: Face protection is required when droplets of blood or other potentially infectious materials may splash or spray during a procedure. When indicated, appropriate covering of the face is important to ensure that the eyes, nose, and mouth are shielded.

3. *PPE*: This includes gowns, aprons, lab coats, clinic jackets, and other similar garments whose use by personnel would depend upon the task and degree of exposure anticipated.

4. *Respiratory equipment*: Mouthpieces, resuscitation bags, and other ventilatory equipment must be made available so that mouth-to-mouth resuscitation is avoided. Reusable mouthpieces must be cleaned and disinfected prior to reuse.

A list of recommended uses of PPE is found on the next page.

Examples of Recommended Personal Protective Equipment for Worker Protection against HIV and HBV Transmission

Task/Activity	Disposable Gloves	Gowns	Masks	Protective Eye Wear
Measuring blood pressure, temperature	No	No	No	No
Bleeding control (spurting blood)	Yes	Yes	Yes	Yes
Bleeding control (minimal bleeding)	Yes	No	No	No
Handling and cleaning contaminated instruments	Yes	No, unless soiling is likely	No	No
Emergency childbirth	Yes	Yes	Yes, if splashing is likely	Yes, if splashing is likely
Endotracheal intubation, esophageal obturator use	Yes	No	No, unless splashing is likely	No, unless splashing is likely
Oral/nasal suctioning, manually cleaning airway	Yes	No	No, unless splashing is likely	No, unless splashing is likely

Source: Reprinted from Department of Health and Human Services (NIOSH) Centers for Disease Control, Guidelines for Prevention of Transmission of HIV and HBV to Health Care and Public Safety Workers, HHS Publications No. 89-107, 1987, Table 4, p. 28.

I. Hepatitis B vaccine

The hepatitis B vaccine is a safe, effective measure available to protect healthcare workers from hepatitis B infection. This vaccination series is offered free of charge. You are not required to receive the vaccine; however, if you choose not to accept this offer you must sign a form declining the vaccination series. Even though you may decide to decline the offer at this time, you still will be able to request the vaccination series at any future date. The hepatitis B vaccine is a noninfectious vaccine prepared genetically from yeast cultures rather than human blood or plasma; therefore, there is no risk of contamination from other bloodborne pathogens nor is there any chance of developing hepatitis B from the vaccine. The vaccination series consists of three separate injections given over a six-month time period. The second injection is given one month after the first one, and the third injection, which completes the vaccination series, is given six months after the first. It is important for individuals to receive all three injections, as more than 90 percent of those vaccinated according to this recommended dosage schedule will develop immunity to the hepatitis B virus.

J. Appropriate actions to take and persons to contact in an emergency involving blood or other potentially infectious materials

When an exposure incident occurs, the area is to be cleaned and decontaminated as soon as possible.

- Cleaning and decontaminating a blood spill

    The appropriate use of PPE must be followed, and, of course, gloves will always be essential. First, wipe the spill with absorbent paper towels and dispose of them in the appropriate container. Wipe the area with a disinfectant approved by the USDA and EPA (antibacteriocidal, antituberculocidal) or a solution of one part bleach to ten parts water. Dispose of paper towels, remove any PPE, and wash hands thoroughly with soap and water.

- Decontaminating laboratory/treatment room

    Standard cleaning and decontamination procedures apply to patient care areas. You must always wear gloves and use EPA- and USDA-approved disinfectants. If disposable coverings are used on surfaces and objects in the room (e.g., exam tables, work carts), they must be replaced immediately after becoming contaminated. These coverings should be discarded and replaced with clean coverings after use with each patient.

K. Occupational exposure/postexposure evaluation and follow-up

After a needlestick, splash, or other body fluid exposure, the exposed area must be washed immediately. For skin contact, the area should be washed thoroughly with soap and water. For mucous membrane contact, flush the area thoroughly with water or normal saline. You must notify the individual who has been assigned the responsibility for filing these incidents. If this person is not a healthcare professional (e.g., physician, nurse), then the employee must be directed to the designated healthcare professional who will perform the postexposure confidential evaluation and follow-up.

L. Information required by employer to provide to employee filing exposure incident

If an exposure incident occurs, you will be provided with

1. Blood testing for HIV and HBV as soon as possible. If you choose to delay testing, the initial specimen will be held for 90 days, within which you may elect to have HIV testing done; otherwise, the sample is discarded after 90 days. Refusal to have HIV and HBV testing must be documented in your record.

2. Postexposure treatment, when medically indicated.

3. Medical follow-up, including evaluation of illnesses (especially those associated with fevers) that may occur primarily within 6 to 12 weeks after the exposure event.

4. The evaluating healthcare professional's written opinion within 15 days of completion of the evaluation. This written opinion will state that: (a) you have been informed of the results of the evaluation, and (b) you have been informed about any medical conditions resulting from exposure to blood or other potentially infectious materials which require

further evaluation or treatment. All other findings or diagnoses will be discussed only with you and kept confidential in your medical record and are not to be included in the written report (e.g., diagnosis of a positive HIV or HBV blood test).

M. Signs, labels, and/or color-coding to identify potentially infectious materials

Warning labels must be placed on all containers with infectious waste. These include refrigerators and freezers containing blood or other potentially infectious materials and any containers used to store, transport, or ship blood or other potentially infectious materials. Labels displaying the universal biohazard precaution symbol are required. These labels must be fluorescent orange or orange-red with letters or symbols in a contrasting color. They must be put on the containers by any method (e.g., adhesive, wire) that prevents their loss or unintentional removal. For all non-sharp contaminated materials, a red bag may replace the use of the biohazard symbol.

N. Question-and-answer period

The person(s) conducting the training session will provide ample opportunity for each trainee to ask questions that may have been generated during the training program. A sample post-training quiz is provided below.

Bloodborne Pathogens Post-Training Evaluation Quiz

1. Pathogens (disease-producing organisms) can be transmitted by
   a. Blood.
   b. Absorption through membranes of the mouth.
   c. Absorption through membranes of the eyes.
   d. All of the above.
2. Hepatitis B virus (HBV) is
   a. Not a problem for healthcare workers.
   b. Not transmitted efficiently by infected blood.
   c. The virus that causes AIDS.
   d. Highly preventable by hepatitis B vaccination.
3. Human immunodeficiency virus (HIV):
   a. Is highly preventable by hepatitis B vaccination.
   b. Is only transmitted by blood.
   c. Affects the immune system.
   d. Affects only intravenous drug users and either homosexual or bisexual men.
4. Hepatitis B vaccination:
   a. Must be taken by all healthcare employees.
   b. Consists of one injection.
   c. Is genetically prepared from yeast and carries no risk of transmitting hepatitis B virus.
   d. Has a minimal charge for administration to all healthcare employees.

5. Personal protective equipment (PPE):
   a. May be gowns, gloves, and masks.
   b. Is provided by the employer at no cost to the employee.
   c. Must be conveniently located and available.
   d. All the above.

6. Disposal of infectious waste:
   a. Can be done in any container, as long as it closes tightly.
   b. Can be done along with regular noninfectious waste.
   c. Must be labeled with a biohazard label or contained in a red bag.
   d. Must occur at the end of each work shift no matter how much or how little waste there is.

7. Handwashing facilities:
   a. Will always be readily available.
   b. Do not have to be used as long as gloves are worn.
   c. May consist of antiseptic cleansers or towelettes until soap and water are available.
   d. Are only necessary when blood drawing occurs.

8. An exposure incident:
   a. May require medical treatment.
   b. Is reported on the OSHA 200 Injury/Illness Log.
   c. Is evaluated by a healthcare professional.
   d. Must be documented in the employee's medical record.
   e. All of the above.

9. The best way to prevent HBV infection is
   a. Always wear gloves when touching patients.
   b. Receive the HBV vaccination.
   c. Use a mouthpiece during CPR resuscitation.
   d. Receive the HIV vaccination.

10. If a spill of blood occurs:
    a. Leave the area immediately and call housekeeping to clean it up.
    b. Immediately put on the appropriate PPE and wipe up the spill, disinfect the area, and dispose of all infectious waste appropriately.
    c. Immediately clean up using a 1:10 bleach-to-water solution.
    d. Wait until the spill dries before attempting any clean up.

11. If recapping contaminated needles is necessary, you should always use the one-handed scoop method or a recapping device to prevent needlestick injury.
    a. True.
    b. False.

12. Universal Precautions should be observed when working with which group?
    a. Male homosexuals.
    b. Only patients with AIDS.
    c. Drug users.
    d. All patients.

13. Masks and protective eyewear are designated to protect you from
    a. Needlestick injury.
    b. Clothing contamination.
    c. Mucous membrane contact.
    d. All of the above.
14. Clearly marked, puncture-resistant containers should be available to dispose of used needles or other disposal sharps.
    a. True.
    b. False.
15. Which activity can spread HIV or HBV from one person to another outside of work?
    a. Using a toilet.
    b. Giving blood.
    c. Shaking hands.
    d. Having sex.

Answer Key

1. d; 2. d; 3. c; 4. c; 5. d; 6. c; 7. c; 8. d; 9. b; 10. b; 11. a; 12. d; 13. d; 14. a; 15. d.

RECORDKEEPING

Medical Records

Accurate medical records for each employee with Exposure must be maintained. These medical records should include

- Name and social security number of the employee
- Employee's hepatitis B vaccination status, including the dates of all the hepatitis B vaccinations and any medical records relative to the employee's ability to receive the vaccination (e.g., the employee has previously received the complete hepatitis B vaccination series or the vaccine is contraindicated for medical reasons)
- Signed Declination Form (ECP VIII-2) if the employee has declined the hepatitis B vaccination offer

If an exposure incident should occur, the records should include

- Postexposure record, which includes circumstances of the exposure incident, the medical evaluation, testing and treatment, results of examinations, and follow-up
- Healthcare professional's written opinion
- Information provided to the healthcare professional

The employee's medical record must be kept confidential and not disclosed or reported without the employee's written consent to any person within or outside the workplace except as required by this section (e.g., to the evaluating healthcare professional) or as may be required by law. These medical records must be kept for the duration of employment plus 30 years.

Training Records
Training records must include the following information:

- Dates of the training sessions
- Contents or summary of the training session or program
- Names and qualifications of persons conducting the training sessions
- Names and job titles of all persons attending the training sessions

The training records must be maintained for three years from the date on which the training session occurred.

Availability
The employee medical and training records must be provided upon request for examination and copying to the subject employee, to anyone having the written consent of the subject employee, to OSHA, and to the National Institute for Occupational Safety and Health in accordance with the provisions in OSHA standard 29 C.F.R. 1910.20.

Transfer of Records
If a company ceases to do business and there is no successor employer to receive and retain the records, the company or organization must make appropriate arrangements for their retention. The company or organization must determine if the records can be retained within another facility. The Director of NIOSH must be informed of the arrangements that have been made for retention of records.

Recordkeeping Checklist
[ ] Medical records are maintained for each employee with occupational exposure.
Medical records include
    [ ] Name and social security number of the employee
    [ ] Employee's hepatitis B vaccination status, including the dates of all HBV vaccinations and any medical reason for the vaccine being contraindicated
    [ ] Signed declination of HBV vaccination offer, if the employee did so
    [ ] All results of examinations, medical testing, and follow-up procedures
    [ ] Healthcare professional's written opinion/evaluation
    [ ] Medical records are kept confidential and not disclosed or reported without the employee's express written consent
    [ ] Medical records are kept for the duration of employment plus 30 years
Training records include
    [ ] Dates of training sessions
    [ ] Contents/summary of training session

[ ] Name(s)/qualifications of trainer(s)
[ ] Names and job titles/classification of all persons attending training session
[ ] Training records are maintained for three years from the date on which training occurred

Responsible person(s)/party: _____[name]_____

BLOODBORNE PATHOGENS PROGRAM

This OSHA standard was developed to protect employees from possible exposure to HIV, hepatitis, and other bloodborne diseases that the employee may be exposed to in the workplace. Additionally, this standard requires the proper labeling and disposal of all medical waste, which includes a wide range of items from a blood sample to a used bandage. Below is the recommended procedure for the development of a program to achieve compliance with this standard.

1. Acquire a copy of the OSHA standard.
2. Review this standard with your management team and acquire management commitment and appropriate funding for the program.
3. Develop a written program incorporating all required elements of the standard which include, but are not limited to, Universal Precautions, Engineering and Work Practice Controls, personal protective equipment, housekeeping, infectious waste disposal, laundry procedures, training requirements, hepatitis B vaccinations, information to be provided to the physician, medical recordkeeping, signs and labels, and availability of medical records.
4. In regard to Universal Precautions, your operations should be analyzed to provide all necessary safeguards to employees who may have possible contact with human blood. Engineering Controls and Work Practice Controls should be examined and evaluated. Procedures should be established for the safe disposal of used needles, used personal protective equipment, and other equipment. Such areas as the refrigerator, cabinets, or freezers where blood could be stored should be prohibited for the storage of food or drink (e.g., employees keeping their lunches in the nurse's refrigerator). Employees should be properly trained in the requirements of this standard. Employees must be prohibited from eating, drinking, smoking, applying cosmetics or lip balm, or handling contact lenses after possible exposure.
5. Where there is the potential for exposure, personal protective equipment such as surgical gloves, gowns, fluid-proof aprons, faceshields, pocket masks, ventilation devices, and so on must be provided to employees. This requirement is especially important for first-aid responders and plant medical personnel who may be exposed when providing first-aid to an injured worker. The personal protective equipment must be appropriately located for easy accessibility. Hypoallergenic personal protective equipment should be made available to employees who may be allergic to the normal personal protective equipment.

6. Employers are required to maintain a clean and sanitary work place. A *written* schedule for cleaning and sanitizing all applicable work areas must be implemented and included in your program. All areas, equipment, and so on that have been exposed (e.g., after an accident) must be cleaned and disinfected. Exposed broken glass must *not* be picked up directly with the hands. A dust broom, vacuum, tongs, or other instrument must be used.

7. All infectious waste, such as bandages, towels, or other items exposed to human blood, must be placed in a closeable, leakproof container or bag with the appropriate label. This waste must be properly disposed of as medical waste according to federal, state, and local regulations. Check with local hospitals or governmental agencies to acquire the regulations applicable to your facility.

8. Contaminated uniforms, smocks, and other items of personal clothing must be laundered in accordance with the standard. *Do not send contaminated uniforms, smocks, etc., to the in-plant laundry.*

9. Appropriate employees should be trained in the requirements of this standard. Employees who may be exposed are required to undergo a medical examination and acquire a hepatitis B vaccination. An example of personnel who may meet this requirement include plant medical personnel and first-aid responders, among others. Training requirements include a copy of the OSHA standard and explanation, in addition to information about the symptoms and epidemiology of bloodborne diseases; transmission of bloodborne pathogens; appropriate methods for recognizing jobs and other activities involving exposure to blood; use and limitations of practices that will prevent or reduce exposure, such as Engineering Controls and personal protective equipment; types, proper use, location, removal, handling, decontamination, or disposal of personal protective equipment; selecting personal protective equipment; hepatitis B vaccine, including information on its efficacy, safety, and benefits; appropriate actions to be taken and persons to contact in an emergency; procedure to be followed if an exposure incident occurs, including the method of reporting, the medical follow-up, and medical counseling; and signs and labels. Additional training may be required in applicable laboratory situations and other circumstances.

10. The employer is required to maintain complete and accurate medical records. These records must include the names and social security numbers of the employees; a record of the employee's hepatitis B vaccination and medical evaluation; the results of all physical examinations, medical testing, and follow-up procedures; the physician's written opinion; and all information provided to the physician. The employer is charged with the responsibility of maintaining the confidentiality of these records.

11. The employer is additionally required to maintain all training records, which must include dates of all training, names of persons conducting the training, and names of all participants. The training records must be maintained for a minimum of five (5) years. Employees and OSHA must be provided the ability to view and copy these records upon request.

12. All containers containing infectious waste, including but not limited to, refrigerators and freezers containing infectious waste, medical disposal containers, and all other containers, *must* be properly labeled. The required label must be fluorescent orange or orange-red with the appropriate symbol and the word "BIOHAZARD" in a contrasting color. This label *must* be affixed to all containers containing infectious waste and must remain affixed until the waste is properly disposed of.

Note: This important standard applies to all employers who have personnel assigned specifically to render first aid or other medical services.

11 Emergency and Disaster Response Hazards

Disasters occur somewhere in the world every day. Simply by watching the evening news or reading a newspaper we can readily find news of disasters hitting individuals, companies, and entire countries. Disasters take various forms, ranging from natural disasters such as tornadoes to man-made disasters such as workplace violence, which happens on a far too frequent basis. No matter what type of disaster, the results are typically the same—substantial loss of life, money, assets, and productivity.

Today's disaster risk has substantially evolved to include areas far beyond the natural disasters of the past, such as cyber-terrorism, product tampering, biological threats, and ecological terrorism, which were virtually unheard of a few short years ago. These disasters can be just as devastating to an organization as a natural disaster; however, the prevention and proactive measures taken are substantially different. Today's safety professional is faced with myriad new and different issues and reactions, ranging from control of the media to shareholder reaction, which were not even considered in the disaster preparedness programs of the past. Because the world is changing, technology is changing, and risks are increasing and evolving, the safety profession must adapt in order to prevent potential disasters from happening where possible, minimize the risks where prevention is not possible, and appropriately react to keep damages to a minimum.

Appropriate planning and preparedness before a disaster occurs are essential to minimizing risks and resulting damages. Individuals involved in disaster preparedness efforts must be appropriately selected and trained so their responses and decision-making during a time of crisis are appropriate. Risks that cannot be appropriately managed internally should be assessed for the possibility of providing external protection or shifting the risk, by means of insurance, for example. Reaction after the disaster must be sure and coordinated in order to minimize damage and avoid causing further harm to the remaining assets, both tangible and intangible.

In this chapter, we have designed a new and innovative method for preparing companies and organizations to address the substantial risk of disasters in the workplace. Our methodology not only encompasses tried-and-true proactive methodology utilized by safety professionals for decades to address natural disasters, but also addresses the often overlooked reactive and postdisaster phases. The progress of society and the accompanying technological changes and terrorist activities require that safety professionals rethink the standard modus operandi for disaster preparedness and expand their proactive and reactive measures to safeguard their company or organization's assets on all levels and at all times. In the blink of an eye, one disaster can decimate years of effort, creativity, and sweat, as well as the monetary and physical assets of a company or organization. Safety professionals are charged with the responsibility of identifying these risks, preparing to protect against these risks, and reacting properly if this risk should develop.

RISK ASSESSMENT

Risks of varying types and magnitudes exist in every workplace on a daily basis; however, some risks are far greater than others and can be disastrous if not identified and properly addressed in terms of reducing the probability of the risk (where feasible), protecting assets through shifting all or a percentage of the risk, and minimizing the potential harm of the risk in a disaster event. The initial step in any disaster preparedness endeavor is to identify the various potential risks, assess the viability and potential of the risk, evaluate the actual probability of the risk, and appraise the potential damage. The potential risks will vary from operation to operation, facility to facility, and location to location. It is vitally important that safety professionals properly identify and assess each facility and operation on an individual basis and customize the preparedness program and efforts to meet the needs of the unique facility. There is no one basic emergency and disaster plan that fits all facilities and operations.

PLANNING DESIGN

Safety professionals should consider a plan that addresses all potential risks and provides the maximum protections for their employees. The plan should include, but not be limited to, the following considerations:

1. Evacuation routes—internal and external
2. Communication systems
3. Command areas
4. Specific responsibilities
5. Triage areas
6. Press areas
7. Liaisons with public sector assistance organizations (i.e., fire, police, EMS)
8. Liaisons with local medical facilities
9. Salvage and security

MEDIA RELATIONS

One essential area often overlooked in the preparation of an emergency and disaster preparedness plan is control of the information and image of your organization being projected to the world by the media. Preplanning in regard to the who, what, when, where, and how of information flow is essential to ensure the accuracy of the information being disseminated about your company and the emergency situation it is facing, as well as creating a favorable opinion among the public about your company's handling of the situation.

Consider the following example. Suppose a publicly held company experiences an explosion that results in ten fatalities, a large number of injured workers, and extensive damage to the facility. After the company notifies the fire department, EMS, and local law enforcement, the local media (who is usually monitoring radio transmissions) dispatches a reporter or television crew to the scene. The television crew will be working under a deadline to provide videotape and information about the incident as quickly as possible. They will try to make the videotape as graphic as possible to interest the viewers and will seek remarks from bystanders, employees, firefighters, or whomever is available (the information provided by witnesses may or may not be correct; however, the television station, for legal reasons, can add a disclaimer at the end of the telecast).

The information gathered at the incident scene will very quickly be picked up by global television networks such as CNN and various newspapers and magazines, in addition to finding its way onto the Internet for worldwide distribution. The information may be slanted or otherwise "editorialized" and often, as when children pass a secret from one to the next, begins to be modified, expanded, and otherwise changed to enhance the story. The facts and truth of the situation are often lost in the shuffle.

Members of the public, sitting at home watching the news or reading the newspaper, will make a value judgment about the company. They may be stockholders, potential employees, or customers who are concerned about the effect this incident will have on trading of the company's stock, future employment with the company, or availability of the company's products. In essence, the information provided to the general public by the media contributes to the opinions formed by the general public about a particular company or organization (whether good or bad) which will affect the public's interactions with the company or organization in the future.

Control of the flow of information after a disaster situation is essential and should be part of an overall comprehensive emergency and disaster preparedness plan. Controlling information about a disaster situation is just as important today as every other phase of the plan due to the potential long-term, downstream, detrimental effects of mishandling the flow of information. As with all other components of the emergency and disaster preparedness plan, appropriate attention should be paid to all areas of the information flow to ensure that a company is putting the best "spin" on an already bad situation. Remember, a bell once sounded cannot be unrung!

As part of the overall emergency and disaster preparedness efforts, consider the following:

- Where will the media acquire their information?
- Will information be screened by legal counsel before providing to the media?
- Will information provided to the media be scripted or ad lib?
- Who will be providing information to the media?
- Is the person providing the information a good representative of your company?
- What is the background of this person?
- What image is projected by this person (e.g., dress, voice quality)?

- What does the media know about your company or organization besides the disaster situation?
- Will the media detrimentally affect the emergency efforts?
- What type of videotape footage will the media be able to acquire?
- What are possible timetables for the media (e.g., a television crew needing to leave the scene by 4:45 p.m. to make a 6:00 p.m. newscast)?
- Where will the media park their vehicles?

Control of information is essential after a disaster situation. Consider the following measures to address the various media-related issues posed above as part of the overall emergency and disaster preparedness plan:

- Identify an area in the parking lot or away from the flow of emergency traffic to which security or management should direct all media vehicles.
- Guide the media to appropriate areas to acquire video footage.
- Maintain security in the media area to prohibit media representatives from "wandering" into the emergency area.
- Designate a selected member of management as spokesperson for the company; no other members of management will talk with the media.
- Consider the dress, tone of voice, ability to remain calm, and other attributes when selecting a spokesperson.
- Provide the spokesperson with an appropriate platform, microphone, and backdrop (e.g., visible company logo) from which to provide information to the media.
- Distribute informational packets about the company to the media.
- Screen all information by legal counsel prior to presentation and keep questions from the media to a minimum.
- Keep in mind the media's deadlines and, if possible, provide information in a timely manner; if information is not provided by the company, hearsay and file footage could be used.
- Always provide truthful information or no information at all.
- Remember the families of the injured or killed employees; do not release names prior to notification of next of kin.
- Remember that the company will be news only for a short period of time. There will be a lot of news potential at the facility or operation immediately after a disaster, but while the story may be front page today, it could be last page tomorrow. It is critical to control the information flow while the situation is most intense.

REMEMBER THE SHAREHOLDER

Safety professionals should also consider the damage inflicted by a disaster on shareholder perceptions of the operations (and, potentially, stock value) of a publicly held company. Companies have become very bottom-line oriented, and the influence of Wall Street on day-to-day operations is significant. A disaster of any magnitude will be immediately communicated by the media throughout the world. Publicly held companies are owned by the individuals and entities holding their stock, and the worth of such a company is intrinsically tied to the value of its shares. If shareholders perceive that a disaster will have a negative impact on the value of their stock, they may decide to sell the stock. If a sufficient number of shares are sold, the value of the shares diminishes which leads to a lower value for the company which equates to tighter budgets, fewer personnel, and other subsequent impacts.

Safety professionals should be aware of how Wall Street can have a positive or negative impact on important aspects of their disaster preparedness program, such as budgets and staffing. Additionally, employees of publicly held companies participate in 401K plans, stock purchasing plans, stock option plans, and other retirement programs and are very interested in the value of their company's stock. Let's examine this issue a little more closely. If a company is "public," the actual owners of the company are the shareholders. The company's board of directors actually works for the shareholders, while the chief executive officer and other upper management officials report to the board of directors. The value of a publicly held company is based upon, in whole or in part, the value of its stock, which is traded daily on one of the stock exchanges (e.g., New York Stock Exchange, American Stock Exchange, NASDAQ).

Today's technology allows individuals to "track" individual stocks on a daily basis on smartphone apps. Additionally, more individual shareholders are managing their portfolios through online trading services, such as E-trade and Ameritrade. In the past, investors may have placed their money in mutual funds and checked the progress once or twice a year, but today they are monitoring the performance of individual company stocks on a daily, or even minute-by-minute, basis. In the past, investors may have "weathered the storm," but today they are more actively involved and more willing to purchase or sell individual stocks at a moment's notice.

So, how does this directly and indirectly affect the overall function of a company? Upper management is now more focused on quarterly returns and looking for a return on investment (ROI) for their shareholders. With such a focus, the safety function has been placed under a microscope to take steps to minimize losses, which may affect the bottom line. The safety manager may be required to report to upper management more often and to collect a wider range of data on a regular basis. Also, given the instantaneous communication available today, a serious accident, chemical release, or other significant event will make the news immediately and could have a detrimental effect upon the individual stock price (for example, the effect of the Valdez disaster on Exxon stock).

When a company is not doing well and shareholders are "jumping ship," the dollars available for safety programs may be reduced. This could have a detrimental effect on management of an emergency situation. Shareholders hold a company's stock to make money; some may hold onto the stock looking for long-run gains, but the current trend is for shareholders to hold stocks for only a short period before trading them. A major disaster situation can quickly place a company's stock into play, with shareholders having to decide whether or not to hold or sell the stock. If the situation appears to be appropriately managed, there is a higher likelihood that shareholders will "weather the storm." If shareholders perceive the situation as being poorly managed, though, they may decide to sell their stock.

In essence, shareholders become a company's bosses. Shareholders normally vote with their feet; if a company experiences a major accident, misses a dividend, or does not meet Wall Street's expectations, there is a substantial likelihood that shareholders will sell off the company's stock. When that happens, the value of the company tends to fall, which affects the company's borrowing power, overall operations, and effectiveness of the safety function.

The key in an emergency situation is to control the flow and type of information being disseminated to the general public. It is important, then, to have a plan established to give shareholders information regarding management of the emergency and how the emergency could affect their investment in the company. The control of information must be managed just as effectively as all other phases of an emergency and disaster plan.

In summary, below are several items to address within an overall disaster preparedness program:

- How will information be provided to the media?
- How will the media be controlled on site?
- Who will act as spokesperson for the company?
- How will the spokesperson be dressed?
- What backdrop will be utilized for videotape and photographs?
- What brokerage house brought the company's stock public?
- What brokerage house represents most of the company's shareholders?
- What lines of communications have been established with these brokerage houses?
- How quickly can information be provided to shareholders?
- Does the information being provided to shareholders reflect a confident posture?
- What type of follow-up should be provided to shareholders?
- Does the company maintain an Internet site?
- What information will the company provide on the Internet?
- Is the information being provided truthful and complete?
- Is the follow-up timely?

In today's market, it is easier for shareholders to sell stock in a company than to risk losses due to a disaster situation. It is vitally important that careful thought and preparation are given to the image the company will be projecting to the world in 30-second sound bites. Shareholders, just like the general public, can make value judgments about a company from a tiny snippet of information. Appropriate management, preparation of image, and effective communication can minimize the overall and long-term damage to an organization. Safety professionals, then, should address every potential risk when developing their emergency and disaster preparedness programs. Every potential issue—ranging from such traditional elements as evacuation routes and triage areas to newer considerations such as media and shareholder information—should be addressed during the planning stage. Selected individuals who will serve in key capacities within the overall plan should be properly trained and prepared to function under stressful conditions if a disaster situation should arise. Preparation is the key to weathering a disaster situation successfully and minimizing the damage in terms of human and economic impact.

Review Questions

1. What are some hazards associated with responding to emergencies and disasters?
2. What are the key considerations for communication with media in the midst of a disaster or emergency situation?
3. Name three emergency and disaster threats safety professionals must consider today that they did not have to consider 50 years earlier.
4. What are some key business considerations during the emergency planning and response process?

12 Workplace Violence

Violence in the workplace has quickly become a major consideration for safety professionals in order to safeguard their employees and organizational assets. In recent years, incidents of workplace violence, ranging from physical altercations to firearms use, have increased significantly. Proper preparedness to address the potential risk of violence in the workplace is a major area of concern for the Occupational Safety and Health Administration (OSHA) and thus for safety professionals, who should consider taking any and all precautions necessary to keep their employees safe. Remember that employees want to return home from work at the end of the day in the same condition as when they went in to work. It is the safety professional's job to ensure that this happens.

INDIVIDUAL WARNING SIGNS

Several studies have found that some individuals are prone to violence in the workplace. Safety professionals should be able to identify the signs of potential violence and work to provide methods to address the situation before the employee becomes violent. An individual that commits workplace violence does not have a sign on his or her back that says, "I will commit a violent act in the workplace." That, of course, would be too simple; instead, the signs are subtle, and even the person him or herself does not know that he or she will commit a violent act one day. One event can start that first domino falling and it is difficult to stop all the dominoes from falling once they have started.

An individual who will commit a violent act in the workplace does not fit a particular profile; there are no clear-cut categories. Many experts have carried out research into workplace violence and have found some broad categories that violent employees fit into. Such employees will have worked at their place of employment for a while and had some success in their jobs. They are dedicated employees who come to work on Saturdays or stay late to finish a project. They make work their life; their work determines how they view themselves, and everything in their life is focused around their work schedule.

Then, one day, they may feel that they were passed up for a promotion or raise and may start holding a grudge against a certain person or the company itself. Another scenario is that the company is downsizing and some people have to be laid off or the company was bought by another one and there will be layoffs. A grudge may be held against a certain person whom they think is responsible for the lack of a raise or promotion or for being laid off.

These people may begin to think that someone is out to get them—whoever laid them off or got the promotion instead. They may also be having a problem outside of work that they are trying to deal with. It could be with their spouse, family, or an annoying neighbor, but they are always having a problem and are always talking about it. They may also have a history of violence, whether a fight in a bar or spousal abuse. They may be fascinated with guns, owning several and even carrying a concealed one.

Some experts say that the typical person prone to workplace violence is a white male in his thirties or forties. As discussed previously, he would have worked for his company for a long period of time, would have a history of violence, and may own a firearm. He is most likely a loner and may or may not have a family; if he does have a family, they will be a source of problems.

Such an employee may try to intimidate the person against whom he holds a grudge by making veiled threats or stalking them. Or, the person may show no outward visible signs of hostility. Take the case of Larry, a technician for an electronics manufacturer in California. Larry was one of several employees laid off due to the recession in 1991. After he was laid off, he kept returning to his former place of employment, seeking advice about finding another job or just visiting with people. Then one day he walked in with a bandolier across his chest and a shotgun in his hand. He shot out the switchboard before the receptionist could call for help. After setting small fires and setting off several pipe bombs, he then calmly walked upstairs and shot and killed a vice president and a regional sales manager. Another executive escaped with his life by hiding under a desk. When Larry was done, he walked out and got on his bicycle and rode away.*

There is no distinct category of people who commit violent acts, but there are warning signals, such as those that were mentioned previously. Noticing some of these warning signals should heighten concern among an employer. An employee displaying some of these signs does not mean he or she will perform a violent act, but the company should have a prevention program in place so when the signs are recognized the program can be put into practice.

The following list includes some of the warning signs for potential for violence in the workplace:

- The person holds grudges against coworkers or the company itself; he might not have received a promotion or a raise that he felt was deserved.
- The person exhibits paranoid behavior; he may feel that the whole world is out to get him.
- The person regularly has problems, which may be related to home, work, or outside friends; the person talks frequently about the problem he is currently dealing with.
- The person has a fascination with firearms and may carry one on his person; he may have a collection and frequently talks about them.
- The person suspects that he is going to be fired or laid off.
- The person is a loner at work and does not mix with coworkers, except maybe for a romantic interest that may lead to the coworker feeling threatened.
- The person may intimidate others by verbal or physical means (e.g., anonymous notes that threaten someone or harassing phone calls).
- The person has difficulty accepting criticism or authority.

---

* Barrier, M., The enemy within, *Nation's Business*, pp. 18–22, February 1995.

- The person likes to push authority to its limit to see how far he can go.
- The person has a history of violent behavior, either before or after employment with the company.
- The person has a history of substance abuse, either drugs or alcohol, or both.
- The person is fascinated by acts of workplace violence and likes to talk about them with coworkers.
- The person may come from an unstable or dysfunctional family.
- Workplace events, such as downsizing or a corporate takeover, may generate stress in the person.
- Activities that expose the person to outside factors may cause an act of violence.*

WORKPLACE VIOLENCE PREVENTION PROGRAM

The Occupational Safety and Health Administration has stated in their regulations [29 U.S.C. §654(a)(1)] that "any employer in the business to provide goods and services that employs people shall provide a safe and healthful work environment." This means that when we go to work we should not have to worry about someone bothering us or harassing us. But that is not always the case anymore, as a disgruntled worker at any time can take out his or her misguided anger on coworkers or the company he or she works for. Such violence has escalated to the point where it is almost out of hand; if companies do not take the initiative to stop it, there is no telling where it could lead.

We all like to count on our workplace being violence-free. Civility and cooperation are basic and essential to the work environment, but now violence is not just on the streets; it has ventured into the office and we must all confront it. In the past, workers tried to leave their home lives at home. We can all remember a boss saying, "Don't bring your personal life to the office." Our world is becoming more violent, and more people are bringing that violence to work with them rather than leaving it at the door. When it happens, it is a rude awakening, as most people think that such violence couldn't happen to them or their employers, it could only happen to someone else.

Many companies have realized what is happening and are taking preventive measures to guard against it. Employee assistance programs (EAPs), which originally started with World War II to deal with alcoholism, are being used to deal with employees who have such problems as marital difficulties or substance abuse. Such programs foster loyalty to a company and a more productive workforce, in addition to assuring employees that the company cares about them, which can sometimes mean more than a raise. Most EAPs are run by outside agencies, as rarely are there people in the company trained to deal with these types of problems, plus confidentiality is ensured.

---

* Barrier, M., Johnson, D., Kiehlbaugh, J., and Kurutz, J., Workplace violence scenario for supervisors, *Human Resources Magazine*, pp. 63–67, 1995.

A preventive maintenance plan for violence should be in effect to deal with violence or even the threat of violence. The following is an outline of a preventive program for a company:

1. Create a management team to develop, review, and implement policies dealing with violence in the workplace. These policies should be included with the company's other policies and procedures. Management backing is very important to any type of employment program. When creating the management team, each team member should be assigned a well-defined responsibility. Once the team has been created, the first step should be an assessment of the possibility of a violent act occurring in their workplace. The team members should make themselves experts on the subject of workplace violence. They should contact all types of outside resources so as to better equip themselves on the subject. The team should then prepare for the possibility of a violent act occurring in their workplace by developing an action plan to deal with a violent act and educating the rest of the employees, supervisors, or team leaders.

2. Create a program that helps employees to detect early warning signals that could ultimately lead to a violent act. This program will enable supervisors or mangers to detect early warning signals and deal with them head on before violence erupts. This program can be incorporated into an already existing training program or be provided as a separate program. Managers, supervisors, or team leaders should be trained to spot the signs, including those discussed previously, and to report them to a member of the management team. The appropriate member of the management team should then conduct an investigation. Any downsizing in the company should put supervisors especially on the alert for any signs of potentially violent behavior.

3. Develop an investigation format for dealing with complaints about a potentially violent employee. The first rule regarding a threat is to take it seriously, as any threat could lead to a violent act. Investigating and recording incidents are key to preventing a violent act in the workplace. Such investigations should include these four Ws:

    a. Who: Who made the actual complaint or threat? Who witnessed the actual threat?

    b. When: When did the threat take place? Are there any events leading up to the actual threat? If so, when did they take place?

    c. What: What exactly happened? Include all facts that pertain to the actual threat and the investigator's assessment of what occurred.

    d. Where: Where did the threat take place? In an office or outside the company? (This is an important aspect that easily gets left out.)

    An incident report form already prepared by the investigator would be the easiest way to handle the investigation. Questions on the report form may include the following:

    • Who made the threat?

    • Who was the threat made against?

    • What were the actual words spoken in the threat?

- Was there any threat of physical harm? If so, what specifically?
- Where did the threat take place (actual physical location)?
- What time did the threat take place (be specific)?
- Who witnessed the actual threat?
- What did the witnesses say occurred (in their own words)? Ask witnesses to write down their statements and attach their statements to the document.
- Is there any other information that might pertain to the incident?

These questions are not inclusive, and a company needs to make their own incident form specific to their organization. It is important when interviewing the person who made the threat to do so in a nonthreatening environment. Having a strategy to deal with the person's anger is important. Focusing only on the facts, not on personal opinions or personality traits, will help the person deal with the situation in a calm, rational manner and control emotional reactions. In addition, there are steps that the interviewer can take to let the person know that the interviewer is interested in what is being said, such as making eye contact; paying full attention to what the person is saying (by avoiding working on notes or answering the phone, as doing so can make the situation worse); asking open-ended questions; not interrupting the person; letting the person finish what he or she is saying; speaking in a calm, rational voice; and repeating what the person says to clarify the situation.

4. Take any action necessary that can prevent an incident from occurring. This may be disciplinary action, talking to a member of the management team in charge of workplace violence, or referral to the company's employee assistance program for counseling. The investigator interviewing the person will come to a set of conclusions. The action that is taken should be based on the conclusions drawn from the interview. In some cases, the person who made the threats may feel better just after talking to the interviewer. This may resolve the issue, but in most cases it won't and further action will be necessary. It is important to take a zero-tolerance policy toward any threat. Consider a company that disciplines someone who makes a threat by making him or her take a day off work without pay. When this same person later hits another employee, he or she is made to take three days off without pay. This type of action will not help the situation; it will only make it worse. The idea is to help an adult solve a problem rather than treat him or her like a naughty child who needs to be disciplined.

5. Develop or improve upon security measures. Technology has advanced security measures. Gone are the days of a security guard walking a solitary route around a building. Instead, we now have security cameras and laser beams to detect intruders. Electronic systems can provide or deny access to persons entering a workplace. Every employer needs to assess their needs and provide adequate security measures. Security companies will provide a review and recommend security measures, some of which will handle every aspect of installing and maintaining the security measures. Providing security is key to protecting employees from the threat of a violent employee and from other types of incidents that could threaten a

company. Immediate improvements that can provide a measure of safety include outdoor lighting, closed-circuit television, intercom systems, and an alarm system. In addition to physical security measures, policies concerning security measures need to be included within the corporate policies and procedures manual. What does the policy state about an employee wanting to work late? Does another employee need to stay also? Are employees even allowed to stay after hours? The policy also needs to address security when a threat has been made. Is security tightened? If so, how? Local law enforcement agencies can be contacted for valuable information regarding threats occurring to other companies, mistakes that were made, and lessons learned from these mistakes. A mistake learned from one company may save a life in another. Local law enforcement can be a vital source of information, if companies utilize them.

6. Develop a crisis plan in the event of an actual violent act in the workplace. The chain of command in the management team needs to be detailed so everyone understands his or her individual responsibilities and there is not any confusion when an incident occurs. When an incident occurs, responsibility shifts from the office manager or supervisor to a management team member, and these responsibilities need to be detailed. When are other agencies notified? The local police should be the first agency notified. A person from the employee assistance program needs to be notified and on hand. Other agencies should be notified as necessary, such as the fire department and local emergency services. In the crisis plan, other resources should be made available, such as a trauma consultant, either a counselor or therapist to assist employees in dealing with the crisis, a security consultant, an in-house legal representative, and a medical physician. What procedures should be put into place immediately to determine how safe the workplace is? Is there an internal security team that can isolate the area until the police department arrives? An immediate assessment is necessary to determine the effect the incident has had on the workplace and an investigation into the incident should be initiated:

  a. What are the facts of the case?
  b. Who were the witnesses?
  c. How can the investigation the employer is conducting be coordinated with the investigation the police department will conduct?
  d. What limits should the employer place on the release of information to other employees? Procedures about releasing information to other employees should be in place, as well as regulations about what can be said during emergency situations.
  e. What are the immediate public relations concerns? How much information will be released to the public and who will release the information?
  f. How does a company preserve control of the day-to-day running of the business while an incident is occurring?

These are just a few of the considerations to be covered by a crisis plan. Each company needs to develop a plan specific to the organization. After a plan has been written, a hypothetical scenario should be enacted to let the employees know that the company is serious about dealing with workplace violence.

7. Use the courts to deal with threats of violence. The company needs to have a zero-tolerance policy toward threats of workplace violence or an actual incident of violence. When an employee has been discharged or is no longer employed by the company and makes a threat against the company or a representative of the company, the use of the court system to deal with the threat can be effective. Restraining orders prohibit a person from seeking access to the person or persons they have threatened. This is not a guarantee that an incident will be prevented, but it is a preventive measure that will lessen the threat of an attack. When a state allows a restraining order, the person making the threat has to stay a certain distance away from the person they made a threat against. Such a restraining order gives the police the authority to make an arrest if the restraining order is violated.

In summation, violence in the workplace is not a traditional area of concern for most safety professionals; however, given the increase in incidents, the potential risk to the organization, and the expertise and skill level of the safety professional, this has become a major area of responsibility for safety professionals. Proper preparation, "thinking outside of the box," and addressing situations before violence occurs can safeguard an organization from the risk of violence. Additionally, proper preparation can minimize the potential damage if an incident of workplace violence occurs. Violence in the workplace is on the rise—safety professionals must be prepared.

Review Questions

1. What are some warning signs that an individual could pose a workplace violence threat?
2. What are the key elements of a workplace violence prevention program?
3. In what ways is communication a critical component of a workplace violence prevention program?
4. What are key topics that should be included in any workplace violence prevention and response training?

Appendix A: Critical Components of a Job Safety Analysis System

The following outline represents key requirements for a corporate policy/system on job safety analysis.

1. Purpose
2. Scope
3. Responsibilities
4. Initiation of JSAs (When must JSAs be completed.)
   a. New Job
   b. Process/Equipment Redesign
   c. Post-Incident/Post–Near-Miss
   d. Periodic
5. Instructions for Conducting JSAs
6. Hazard Assessment/Risk Assessment Process (Students should not simply refer to another policy.)
   a. Risk Matrices
   b. Risk Assessment Codes, Values, and Descriptions/Definitions
   c. Quantification of Risk
   d. Risk Management Prioritization
7. Hazard Control Requirements
   a. Hierarchy of Controls
   b. Means to Assess and Prescribe Hazard Controls
   c. Follow-Up/Corrective Action
8. Training Requirements
9. JSA Documentation Requirements
10. Review and Approval of Completed JSAs
11. Periodic Review of JSA Program Performance
12. Forms and Instructions

Appendix B: Employee Workplace Rights

INTRODUCTION

The Occupational Safety and Health (OSH) Act of 1970 created the Occupational Safety and Health Administration (OSHA) within the Department of Labor and encouraged employers and employees to reduce workplace hazards and to implement safety and health programs. In so doing, this gave employees many new rights and responsibilities, including the right to do the following:

- Review copies of appropriate standards, rules, regulations, and requirements that the employer should have available at the workplace.
- Request information from the employer on safety and health hazards in the workplace, precautions that may be taken, and procedures to be followed if the employee is involved in an accident or is exposed to toxic substances.
- Have access to relevant employee exposure and medical records.
- Request the OSHA area director to conduct an inspection if he or she believes hazardous conditions or violations of standards exist in the workplace.
- Have an authorized employee representative accompany the OSHA compliance officer during the inspection tour.
- Respond to questions from the OSHA compliance officer, particularly if there is no authorized employee representative accompanying the compliance officer on the inspection "walkaround."
- Observe any monitoring or measuring of hazardous materials and see the resulting records, as specified under the OSH Act, and as required by OSHA standards.
- Have an authorized representative, or themselves, review the Log and Summary of Occupational Injuries (OSHA 300 log and 300A Summary) at a reasonable time and in a reasonable manner.
- Object to the abatement period set by OSHA for correcting any violation in the citation issued to the employer by writing to the OSHA area director within 15 working days from the date the employer receives the citation.
- Submit a written request to the National Institute for Occupational Safety and Health (NIOSH) for information on whether any substance in the workplace has potentially toxic effects in the concentration being used, and have their names withheld from the employer, if requested.
- Be notified by an employer if the employer applies for a variance from an OSHA standard, testify at a variance hearing, and appeal the final decision.

- Have their names withheld from their employer, upon request to OSHA, if they sign and file written complaints.
- Be advised of OSHA actions regarding a complaint and request an informal review of any decision not to inspect or to issue a citation.
- File a Section 11(c) discrimination complaint if punished for exercising the above rights or for refusing to work when faced with imminent danger of death or serious injury and there is insufficient time for OSHA to inspect.

Pursuant to Section 18 of the act, states can develop and operate their own occupational safety and health programs under state plans approved and monitored by federal OSHA. States that assume responsibility for their own occupational safety and health programs must have provisions at least as effective as those of federal OSHA, including the protection of employee rights. There are currently 25 state plans. Twenty-one states and two territories administer plans covering both private and state and local government employment, and two states cover only the public sector. All the rights and responsibilities described in this appendix are similarly provided by state programs. Any interested person or groups of persons, including employees, who have a complaint concerning the operation or administration of a state plan may submit a Complaint About State Program Administration (CASPA) to the appropriate OSHA regional administrator. Under CASPA procedures, the OSHA regional administrator investigates these complaints and informs the state and the complainant of these findings. Corrective action is recommended when required.

OSHA STANDARDS AND WORKPLACE HAZARDS

Before OSHA issues, amends, or deletes regulations, the agency publishes them in the *Federal Register* so that interested persons or groups may comment. The employer has a legal obligation to inform employees of OSHA safety and health standards that apply to their workplace. Upon request, the employer must make available copies of those standards and the OSHA law itself. If more information is needed about workplace hazards than the employer can supply, it can be obtained from the nearest OSHA area office.

Under the OSH Act, employers have a general duty to provide work and a workplace free from recognized hazards. Citations may be issued by OSHA when violations of standards are found and for violations of the general duty clause, even if no OSHA standard applies to the particular hazard. The employer also must display in a prominent place the official OSHA poster that describes rights and responsibilities under the OSH Act.

ACCESS TO EXPOSURE AND MEDICAL RECORDS

Employers must inform employees of the existence, location, and availability of their medical and exposure records when employees first begin employment and at least annually thereafter. Employers also must provide these records to employees or their designated representatives, upon request. Whenever an employer plans to stop doing business and there is no successor employer to receive and maintain these records, the employer must notify employees of their right of access to records at least three months before the employer ceases to do business. OSHA standards require the employer to measure exposure to harmful substances; the employee (or representative) has the right to observe the testing and to examine the records of the results. If the exposure levels are above the limit set by the standard, the employer must tell employees what will be done to reduce the exposure.

OSHA INSPECTIONS

If a hazard is not being corrected, an employee should contact the OSHA area office (or state program office) having jurisdiction. If the employee submits a written complaint and the OSHA area or state office determines that there are reasonable grounds for believing that a violation or danger exists, the office conducts an inspection.

EMPLOYEE REPRESENTATIVE

Under Section 8(e), the workers' representative has a right to accompany an OSHA compliance officer (also referred to as a compliance safety and health officer, CSHO, or inspector) during an inspection. The representative must be chosen by the union (if there is one) or by the employees. Under no circumstances may the employer choose the workers' representative.

If employees are represented by more than one union, each union may choose a representative. Normally, the representative of each union will not accompany the inspector for the entire inspection but will join the inspection only when it reaches the area where those union members work.

An OSHA inspector may conduct a comprehensive inspection of the entire workplace or a partial inspection limited to certain areas or aspects of the operation.

HELPING THE COMPLIANCE OFFICER

Workers have a right to talk privately to the compliance officer on a confidential basis whether or not a workers' representative has been chosen. Workers are encouraged to point out hazards, describe accidents or illnesses that resulted from those hazards, describe past worker complaints about hazards, and inform the inspector if working conditions are not normal during the inspection.

OBSERVING MONITORING

If health hazards are present in the workplace, a special OSHA health inspection may be conducted by an industrial hygienist. This OSHA inspector may take samples to measure levels of dust, noise, fumes, or other hazardous materials.

OSHA will inform the employee representative as to whether the employer is in compliance. The inspector also will gather detailed information about the employer's efforts to control health hazards, including results of tests the employer may have conducted.

Reviewing OSHA Injury and Illness Records

If the employer is required to maintain records of work-related injuries and illnesses (OSHA 300 log, 300A, etc.), the employees or their representative have the right to review those records.

AFTER AN INSPECTION

At the end of the inspection, the OSHA inspector will meet with the employer and the employee representatives in a closing conference to discuss the abatement of any hazards that may have been found. If it is not practical to hold a joint conference, separate conferences will be held, and OSHA will provide written summaries, on request.

During the closing conference, the employee representative may describe, if not reported already, what hazards exist, what should be done to correct them, and how long it should take. Other facts about the history of health and safety conditions at the workplace may also be provided.

Challenging Abatement Period

Whether or not the employer accepts OSHA's actions, the employee (or representative) has the right to contest the time OSHA allows for correcting a hazard. This contest must be filed in writing with the OSHA area director within 15 working days after the citation is issued. The contest will be decided by the Occupational Safety and Health Review Commission. The Review Commission is an independent agency and is not part of the Department of Labor.

Variances

Some employers may not be able to comply fully with a new safety and health standard in the time provided due to shortages of personnel, materials, or equipment. In situations like these, employers may apply to OSHA for a temporary variance from the standard. In other cases, employers may be using methods or equipment that differ from those prescribed by OSHA, but which the employer believes are equal to or better than OSHA's requirements and would qualify for consideration as a permanent variance. Applications for a permanent variance must basically contain the same information as those for temporary variances.

The employer must certify that workers have been informed of the variance application, that a copy has been given to the employee's representative, and that a summary of the application has been posted wherever notices are normally posted in the workplace. Employees also must be informed that they have the right to request a hearing on the application.

Employees, employers, and other interested groups are encouraged to participate in the variance process. Notices of variance application are published in the *Federal Register*, inviting all interested parties to comment on the action.

CONFIDENTIALITY

OSHA will not tell the employer who requested the inspection unless the complainant indicates that he or she has no objection.

REVIEW IF NO INSPECTION IS MADE

The OSHA area director evaluates the complaint from the employee or representative and decides whether it is valid. If the area director decides not to inspect the workplace, he or she will send a certified letter to the complainant explaining the decision and the reasons for it. Complainants must be informed that they have the right to request further clarification of the decision from the area director; if still dissatisfied, they can appeal to the OSHA regional administrator for an informal review. Similarly, a decision by an area director not to issue a citation after an inspection is subject to further clarification from the area director and to an informal review by the regional administrator.

DISCRIMINATION FOR USING RIGHTS

Employees have a right to seek safety and health on the job without fear of punishment. That right is spelled out in Section 11(c) of the act. The law says the employer "shall not" punish or discriminate against employees for exercising such rights as complaining to the employer, union, OSHA, or any other government agency about job safety and health hazards; or for participating in OSHA inspections, conferences, hearings, or other OSHA-related activities.

Although there is nothing in the OSHA law that specifically gives an employee the right to refuse to perform an unsafe or unhealthful job assignment, OSHA's regulations, which have been upheld by the U.S. Supreme Court, provide that an employee may refuse to work when faced with an imminent danger of death or serious injury. The conditions necessary to justify a work refusal are very stringent, however, and a work refusal should be an action taken only as a last resort. If time permits, the unhealthful or unsafe condition must be reported to OSHA or other appropriate regulatory agency.

A state that is administering its own occupational safety and health enforcement program pursuant to Section 18 of the act must have provisions as effective as those of Section 11(c) to protect employees from discharge or discrimination. OSHA, however, retains its Section 11(c) authority in all states regardless of the existence of an OSHA-approved state occupational safety and health program.

Workers who believe that they have been punished for exercising safety and health rights must contact the nearest OSHA office within 30 days of the time they learn of the alleged discrimination. A representative of the employee's choosing can file the 11(c) complaint for the worker. Following a complaint, OSHA will contact the complainant and conduct an in-depth interview to determine whether an investigation is necessary.

If evidence supports the conclusion that the employee has been punished for exercising safety and health rights, OSHA will ask the employer to restore that worker's job, earnings, and benefits. If the employer declines to enter into a voluntary settlement, OSHA may take the employer to court. In such cases, an attorney of the Department of Labor will conduct litigation on behalf of the employee to obtain this relief.

Employee Responsibilities

Although OSHA does not cite employees for violations of their responsibilities, each employee "shall comply with all occupational safety and health standards and all rules, regulations, and orders issued under the act" that are applicable. Employee responsibilities and rights in states with their own occupational safety and health programs are generally the same as for workers in states covered by federal OSHA. An employee should do the following:

- Read the OSHA poster at the jobsite.
- Comply with all applicable OSHA standards.
- Follow all lawful employer safety and health rules and regulations, and wear or use prescribed protective equipment while working.
- Report hazardous conditions to the supervisor.
- Report any job-related injury or illness to the employer, and seek treatment promptly.
- Cooperate with the OSHA compliance officer conducting an inspection if he or she inquires about safety and health conditions in the workplace.
- Exercise rights under the act in a responsible manner.

Appendix C: Appendices to OSHA's Respiratory Protection Standard 29 CFR § 1910.134

APPENDIX A TO § 1910.134: FIT TESTING PROCEDURES (MANDATORY)

PART I. OSHA-ACCEPTED FIT TEST PROTOCOLS

A. Fit testing procedures—general requirements.

The employer shall conduct fit testing using the following procedures. The requirements in this appendix apply to all OSHA-accepted fit test methods, both QLFT and QNFT.

1. The test subject shall be allowed to pick the most acceptable respirator from a sufficient number of respirator models and sizes so that the respirator is acceptable to, and correctly fits, the user.

2. Prior to the selection process, the test subject shall be shown how to put on a respirator, how it should be positioned on the face, how to set strap tension, and how to determine an acceptable fit. A mirror shall be available to assist the subject in evaluating the fit and positioning of the respirator. This instruction may not constitute the subject's formal training on respirator use, because it is only a review.

3. The test subject shall be informed that he or she is being asked to select the respirator that provides the most acceptable fit. Each respirator represents a different size and shape, and if fitted and used properly, will provide adequate protection.

4. The test subject shall be instructed to hold each chosen facepiece up to the face and eliminate those that obviously do not give an acceptable fit.

5. The more acceptable facepieces are noted in case the one selected proves unacceptable; the most comfortable mask is donned and worn at least five minutes to assess comfort. Assistance in assessing comfort can be given by discussing the points in the following item A.6. If the test subject is not familiar with using a particular respirator, the test subject shall be directed to don the mask several times and to adjust the straps each time to become adept at setting proper tension on the straps.

6. Assessment of comfort shall include a review of the following points with the test subject and allowing the test subject adequate time to determine the comfort of the respirator:
   a. Position of the mask on the nose
   b. Room for eye protection
   c. Room to talk
   d. Position of mask on face and cheeks
7. The following criteria shall be used to help determine the adequacy of the respirator fit:
   a. Chin properly placed
   b. Adequate strap tension, not overly tightened
   c. Fit across nose bridge
   d. Respirator of proper size to span distance from nose to chin
   e. Tendency of respirator to slip
   f. Self-observation in mirror to evaluate fit and respirator position.
8. The test subject shall conduct a user seal check, either the negative and positive pressure seal checks described in Appendix B-1 of this section or those recommended by the respirator manufacturer which provide equivalent protection to the procedures in Appendix B-1. Before conducting the negative and positive pressure checks, the subject shall be told to seat the mask on the face by moving the head from side to side and up and down slowly while taking in a few slow, deep breaths. Another facepiece shall be selected and retested if the test subject fails the user seal check tests.
9. The test shall not be conducted if there is any hair growth between the skin and the facepiece sealing surface, such as stubble beard growth, beard, mustache, or sideburns that cross the respirator sealing surface. Any type of apparel that interferes with a satisfactory fit shall be altered or removed.
10. If a test subject exhibits difficulty in breathing during the tests, she or he shall be referred to a physician or other licensed health care professional, as appropriate, to determine whether the test subject can wear a respirator while performing her or his duties.
11. If the employee finds the fit of the respirator unacceptable, the test subject shall be given the opportunity to select a different respirator and to be retested.
12. Exercise regimen. Prior to the commencement of the fit test, the test subject shall be given a description of the fit test and the test subject's responsibilities during the test procedure. The description of the process shall include a description of the test exercises that the subject will be performing. The respirator to be tested shall be worn for at least five minutes before the start of the fit test.
13. The fit test shall be performed while the test subject is wearing any applicable safety equipment that may be worn during actual respirator use which could interfere with respirator fit.
14. Test Exercises.
   a. Employers must perform the following test exercises for all fit testing methods prescribed in this appendix, except for the CNP quantitative

fit testing protocol and the CNP REDON quantitative fit testing protocol. For these two protocols, employers must ensure that the test subjects (i.e., employees) perform the exercise procedure specified in Part I.C.4(b) of this appendix for the CNP quantitative fit testing protocol, or the exercise procedure described in Part I.C.5(b) of this appendix for the CNP REDON quantitative fit-testing protocol. For the remaining fit testing methods, employers must ensure that employees perform the test exercises in the appropriate test environment in the following manner:

(1) Normal breathing. In a normal standing position, without talking, the subject shall breathe normally.

(2) Deep breathing. In a normal standing position, the subject shall breathe slowly and deeply, taking caution so as not to hyperventilate.

(3) Turning head side to side. Standing in place, the subject shall slowly turn his or her head from side to side between the extreme positions on each side. The head shall be held at each extreme momentarily so the subject can inhale at each side.

(4) Moving head up and down. Standing in place, the subject shall slowly move his or her head up and down. The subject shall be instructed to inhale in the up position (i.e., when looking toward the ceiling).

(5) Talking. The subject shall talk out loud slowly and loud enough so as to be heard clearly by the test conductor. The subject can read from a prepared text such as the Rainbow Passage, count backward from 100, or recite a memorized poem or song.

Rainbow Passage

When the sunlight strikes raindrops in the air, they act like a prism and form a rainbow. The rainbow is a division of white light into many beautiful colors. These take the shape of a long round arch, with its path high above, and its two ends apparently beyond the horizon. There is, according to legend, a boiling pot of gold at one end. People look, but no one ever finds it. When a man looks for something beyond reach, his friends say he is looking for the pot of gold at the end of the rainbow.

(6) Grimace. The test subject shall grimace by smiling or frowning. (This applies only to QNFT testing; it is not performed for QLFT.)

(7) Bending over. The test subject shall bend at the waist as if he or she were to touch his or her toes. Jogging in place shall be substituted for this exercise in those test environments such as shroud type QNFT or QLFT units that do not permit bending over at the waist.

(8) Normal breathing. Same as exercise (1).

b. Each test exercise shall be performed for one minute except for the grimace exercise, which shall be performed for 15 seconds. The test subject shall be questioned by the test conductor regarding the comfort of the respirator upon completion of the protocol. If it has become unacceptable, another model of respirator shall be tried. The respirator shall not be adjusted once the fit test exercises begin. Any adjustment voids the test, and the fit test must be repeated.

B. Qualitative Fit Test (QLFT) protocols.

1. General.
    (a) The employer shall ensure that persons administering QLFT are able to prepare test solutions, calibrate equipment and perform tests properly, recognize invalid tests, and ensure that test equipment is in proper working order.
    (b) The employer shall ensure that QLFT equipment is kept clean and well-maintained so as to operate within the parameters for which it was designed.
2. Isoamyl Acetate Protocol. *Note*: This protocol is not appropriate to use for the fit testing of particulate respirators. If used to fit test particulate respirators, the respirator must be equipped with an organic vapor filter.
    (a) Odor Threshold Screening. Odor threshold screening, performed without wearing a respirator, is intended to determine if the individual tested can detect the odor of isoamyl acetate at low levels.
        (1) Three one-liter glass jars with metal lids are required.
        (2) Odor-free water (e.g., distilled or spring water) at approximately 25C° (77F°) shall be used for the solutions.
        (3) The isoamyl acetate (IAA; also known at isopentyl acetate) stock solution is prepared by adding one milliliter of pure IAA to 800 milliliters of odor-free water in a one liter jar, closing the lid, and shaking for 30 seconds. A new solution shall be prepared at least weekly.
        (4) The screening test shall be conducted in a room separate from the room used for actual fit testing. The two rooms shall be well-ventilated to prevent the odor of IAA from becoming evident in the general room air where testing takes place.
        (5) The odor test solution is prepared in a second jar by placing 0.4 milliliters of the stock solution into 500 milliliters of odor-free water using a clean dropper or pipette. The solution shall be shaken for 30 seconds and allowed to stand for two to three minutes so that the IAA concentration above the liquid may reach equilibrium. This solution shall be used for only one day.
        (6) A test blank shall be prepared in a third jar by adding 500 cc of odor-free water.
        (7) The odor test and test blank jar lids shall be labeled (e.g., 1 and 2) for jar identification. Labels shall be placed on the lids so that they can be peeled off periodically and switched to maintain the integrity of the test.
        (8) The following instruction shall be typed on a card and placed on the table in front of the two test jars (i.e., 1 and 2): "The purpose of this

test is to determine if you can smell banana oil at a low concentration. The two bottles in front of you contain water. One of these bottles also contains a small amount of banana oil. Be sure the covers are on tight, then shake each bottle for two seconds. Unscrew the lid of each bottle, one at a time, and sniff at the mouth of the bottle. Indicate to the test conductor which bottle contains banana oil."

(9) The mixtures used in the IAA odor detection test shall be prepared in an area separate from where the test is performed, in order to prevent olfactory fatigue in the subject.

(10) If the test subject is unable to correctly identify the jar containing the odor test solution, the IAA qualitative fit test shall not be performed.

(11) If the test subject correctly identifies the jar containing the odor test solution, the test subject may proceed to respirator selection and fit testing.

(b) Isoamyl Acetate Fit Test.

(1) The fit test chamber shall be a clear 55-gallon drum liner suspended inverted over a two-foot diameter frame so that the top of the chamber is about six inches above the test subject's head. If no drum liner is available, a similar chamber shall be constructed using plastic sheeting. The inside top center of the chamber shall have a small hook attached.

(2) Each respirator used for the fitting and fit testing shall be equipped with organic vapor cartridges or offer protection against organic vapors.

(3) After selecting, donning, and properly adjusting a respirator, the test subject shall wear it to the fit testing room. This room shall be separate from the room used for odor threshold screening and respirator selection, and shall be well-ventilated, as by an exhaust fan or lab hood, to prevent general room contamination.

(4) A copy of the test exercises and any prepared text from which the subject is to read shall be taped to the inside of the test chamber.

(5) Upon entering the test chamber, the test subject shall be given a six-inch by five-inch piece of paper towel or other porous, absorbent, single-ply material, folded in half and wetted with 0.75 milliliters of pure IAA. The test subject shall hang the wet towel on the hook at the top of the chamber. An IAA test swab or ampule may be substituted for the IAA wetted paper towel provided it has been demonstrated that the alternative IAA source will generate an IAA test atmosphere with a concentration equivalent to that generated by the paper towel method.

(6) Allow two minutes for the IAA test concentration to stabilize before starting the fit test exercises. This would be an appropriate time to talk with the test subject; to explain the fit test, the importance of his or her cooperation, and the purpose for the test exercises; or to demonstrate some of the exercises.

(7) If at any time during the test, the subject detects the banana-like odor of IAA, the test is failed. The subject shall quickly exit from the test chamber and leave the test area to avoid olfactory fatigue.

(8) If the test is failed, the subject shall return to the selection room and remove the respirator. The test subject shall repeat the odor sensitivity test, select and put on another respirator, return to the test area and again begin the fit test procedure described in (b)(1) through (7) above. The process continues until a respirator that fits well has been found. Should the odor sensitivity test be failed, the subject shall wait at least five minutes before retesting. Odor sensitivity will usually have returned by this time.

(9) If the subject passes the test, the efficiency of the test procedure shall be demonstrated by having the subject break the respirator face seal and take a breath before exiting the chamber.

(10) When the test subject leaves the chamber, the subject shall remove the saturated towel and return it to the person conducting the test, so that there is no significant IAA concentration buildup in the chamber during subsequent tests. The used towels shall be kept in a self-sealing plastic bag to keep the test area from being contaminated.

3. Saccharin Solution Aerosol Protocol. The entire screening and testing procedure shall be explained to the test subject prior to the conduct of the screening test.

(a) Taste threshold screening. The saccharin taste threshold screening, performed without wearing a respirator, is intended to determine whether the individual being tested can detect the taste of saccharin.

(1) During threshold screening as well as during fit testing, subjects shall wear an enclosure about the head and shoulders that is approximately 12 inches in diameter by 14 inches tall with at least the front portion clear and that allows free movements of the head when a respirator is worn. An enclosure substantially similar to the 3M hood assembly, parts # FT 14 and # FT 15 combined, is adequate.

(2) The test enclosure shall have a 3/4-inch (1.9 centimeter) hole in front of the test subject's nose and mouth area to accommodate the nebulizer nozzle.

(3) The test subject shall don the test enclosure. Throughout the threshold screening test, the test subject shall breathe through his or her slightly open mouth with tongue extended. The subject is instructed to report when he or she detects a sweet taste.

(4) Using a DeVilbiss Model 40 Inhalation Medication Nebulizer or equivalent, the test conductor shall spray the threshold check solution into the enclosure. The nozzle is directed away from the nose and mouth of the person. This nebulizer shall be clearly marked to distinguish it from the fit test solution nebulizer.

(5) The threshold check solution is prepared by dissolving 0.83 grams of sodium saccharin USP in 100 milliliters of warm water. It can

be prepared by putting 1 milliliter of the fit test solution (see (b)(5) below) in 100 milliliters of distilled water.

(6) To produce the aerosol, the nebulizer bulb is firmly squeezed so that it collapses completely, then released and allowed to fully expand.

(7) Ten squeezes are repeated rapidly and then the test subject is asked whether the saccharin can be tasted. If the test subject reports tasting the sweet taste during the ten squeezes, the screening test is completed. The taste threshold is noted as ten regardless of the number of squeezes actually completed.

(8) If the first response is negative, ten more squeezes are repeated rapidly and the test subject is again asked whether the saccharin is tasted. If the test subject reports tasting the sweet taste during the second ten squeezes, the screening test is completed. The taste threshold is noted as twenty regardless of the number of squeezes actually completed.

(9) If the second response is negative, ten more squeezes are repeated rapidly and the test subject is again asked whether the saccharin is tasted. If the test subject reports tasting the sweet taste during the third set of ten squeezes, the screening test is completed. The taste threshold is noted as thirty regardless of the number of squeezes actually completed.

(10) The test conductor will take note of the number of squeezes required to solicit a taste response.

(11) If the saccharin is not tasted after 30 squeezes (step 10), the test subject is unable to taste saccharin and may not perform the saccharin fit test.

Note to paragraph 3.(a): If the test subject eats or drinks something sweet before the screening test, he or she may be unable to taste the weak saccharin solution.

(12) If a taste response is elicited, the test subject shall be asked to take note of the taste for reference in the fit test.

(13) Correct use of the nebulizer means that approximately 1 milliliter of liquid is used at a time in the nebulizer body.

(14) The nebulizer shall be thoroughly rinsed in water, shaken dry, and refilled at least each morning and afternoon or at least every four hours.

(b) Saccharin solution aerosol fit test procedure.

(1) The test subject may not eat, drink (except plain water), smoke, or chew gum for 15 minutes before the test.

(2) The fit test uses the same enclosure described in 3.(a) above.

(3) The test subject shall don the enclosure while wearing the respirator selected in Section I.A. of this appendix. The respirator shall be properly adjusted and equipped with a particulate filter(s).

(4) A second DeVilbiss Model 40 Inhalation Medication Nebulizer or equivalent is used to spray the fit test solution into the enclosure.

This nebulizer shall be clearly marked to distinguish it from the screening test solution nebulizer.

(5) The fit test solution is prepared by adding 83 grams of sodium saccharin to 100 milliliters of warm water.

(6) As before, the test subject shall breathe through the slightly open mouth with tongue extended, and report if he or she tastes the sweet taste of saccharin.

(7) The nebulizer is inserted into the hole in the front of the enclosure and an initial concentration of saccharin fit test solution is sprayed into the enclosure using the same number of squeezes (either ten, twenty, or thirty squeezes) based on the number of squeezes required to elicit a taste response as noted during the screening test. A minimum of ten squeezes is required.

(8) After generating the aerosol, the test subject shall be instructed to perform the exercises in Section I.A.14 of this appendix.

(9) Every 30 seconds the aerosol concentration shall be replenished using one half the original number of squeezes used initially (e.g., five, ten, or fifteen).

(10) The test subject shall indicate to the test conductor if at any time during the fit test the taste of saccharin is detected. If the test subject does not report tasting the saccharin, the test is passed.

(11) If the taste of saccharin is detected, the fit is deemed unsatisfactory and the test is failed. A different respirator shall be tried and the entire test procedure is repeated (taste threshold screening and fit testing).

(12) Since the nebulizer has a tendency to clog during use, the test operator must make periodic checks of the nebulizer to ensure that it is not clogged. If clogging is found at the end of the test session, the test is invalid.

4. Bitrex™ (Denatonium Benzoate) Solution Aerosol Qualitative Fit Test Protocol. The Bitrex™ (denatonium benzoate) solution aerosol QLFT protocol uses the published saccharin test protocol because that protocol is widely accepted. Bitrex is routinely used as a taste aversion agent in household liquids which children should not be drinking and is endorsed by the American Medical Association, the National Safety Council, and the American Association of Poison Control Centers. The entire screening and testing procedure shall be explained to the test subject prior to the conduct of the screening test.

(a) Taste Threshold Screening. The Bitrex taste threshold screening, performed without wearing a respirator, is intended to determine whether the individual being tested can detect the taste of Bitrex.

(1) During threshold screening as well as during fit testing, subjects shall wear an enclosure about the head and shoulders that is approximately 12 inches (30.5 centimeters) in diameter by 14 inches (35.6 centimeters) tall. The front portion of the enclosure shall be clear from the respirator and allow free movement of the head when a

respirator is worn. An enclosure substantially similar to the 3M hood assembly, parts # FT 14 and # FT 15 combined, is adequate.

(2) The test enclosure shall have a 3/4-inch (1.9 centimeter) hole in front of the test subject's nose and mouth area to accommodate the nebulizer nozzle.

(3) The test subject shall don the test enclosure. Throughout the threshold screening test, the test subject shall breathe through his or her slightly open mouth with tongue extended. The subject is instructed to report when he or she detects a bitter taste.

(4) Using a DeVilbiss Model 40 Inhalation Medication Nebulizer or equivalent, the test conductor shall spray the Threshold Check Solution into the enclosure. This nebulizer shall be clearly marked to distinguish it from the fit test solution nebulizer.

(5) The Threshold Check Solution is prepared by adding 13.5 milligrams of Bitrex to 100 milliliters of 5 percent salt (NaCl) solution in distilled water.

(6) To produce the aerosol, the nebulizer bulb is firmly squeezed so that the bulb collapses completely, and is then released and allowed to fully expand.

(7) An initial ten squeezes are repeated rapidly and then the test subject is asked whether the Bitrex can be tasted. If the test subject reports tasting the bitter taste during the ten squeezes, the screening test is completed. The taste threshold is noted as ten regardless of the number of squeezes actually completed.

(8) If the first response is negative, ten more squeezes are repeated rapidly and the test subject is again asked whether the Bitrex is tasted. If the test subject reports tasting the bitter taste during the second ten squeezes, the screening test is completed. The taste threshold is noted as twenty regardless of the number of squeezes actually completed.

(9) If the second response is negative, ten more squeezes are repeated rapidly and the test subject is again asked whether the Bitrex is tasted. If the test subject reports tasting the bitter taste during the third set of ten squeezes, the screening test is completed. The taste threshold is noted as thirty regardless of the number of squeezes actually completed.

(10) The test conductor will take note of the number of squeezes required to solicit a taste response.

(11) If the Bitrex is not tasted after thirty squeezes (step 10), the test subject is unable to taste Bitrex and may not perform the Bitrex fit test.

(12) If a taste response is elicited, the test subject shall be asked to take note of the taste for reference in the fit test.

(13) Correct use of the nebulizer means that approximately 1 milliliter of liquid is used at a time in the nebulizer body.

(14) The nebulizer shall be thoroughly rinsed in water, shaken to dry, and refilled at least each morning and afternoon or at least every four hours.

(b)   Bitrex Solution Aerosol Fit Test Procedure.

   (1)   The test subject may not eat, drink (except plain water), smoke, or chew gum for 15 minutes before the test.

   (2)   The fit test uses the same enclosure as that described in 4.(a) above.

   (3)   The test subject shall don the enclosure while wearing the respirator selected according to Section I.A. of this appendix. The respirator shall be properly adjusted and equipped with any type particulate filter(s).

   (4)   A second DeVilbiss Model 40 Inhalation Medication Nebulizer or equivalent is used to spray the fit test solution into the enclosure. This nebulizer shall be clearly marked to distinguish it from the screening test solution nebulizer.

   (5)   The fit test solution is prepared by adding 337.5 milligrams of Bitrex to 200 milliliter of a 5 percent salt (NaCl) solution in warm water.

   (6)   As before, the test subject shall breathe through his or her slightly open mouth with tongue extended, and be instructed to report if he or she tastes the bitter taste of Bitrex.

   (7)   The nebulizer is inserted into the hole in the front of the enclosure and an initial concentration of the fit test solution is sprayed into the enclosure using the same number of squeezes (either ten, twenty, or thirty squeezes) based on the number of squeezes required to elicit a taste response as noted during the screening test.

   (8)   After generating the aerosol, the test subject shall be instructed to perform the exercises in Section I.A.14. of this appendix.

   (9)   Every 30 seconds the aerosol concentration shall be replenished using one half the number of squeezes used initially (e.g., five, ten, or fifteen).

   (10)   The test subject shall indicate to the test conductor if at any time during the fit test the taste of Bitrex is detected. If the test subject does not report tasting the Bitrex, the test is passed.

   (11)   If the taste of Bitrex is detected, the fit is deemed unsatisfactory and the test is failed. A different respirator shall be tried and the entire test procedure is repeated (taste threshold screening and fit testing).

5. Irritant Smoke (Stannic Chloride) Protocol. This qualitative fit test uses a person's response to the irritating chemicals released in the "smoke" produced by a stannic chloride ventilation smoke tube to detect leakage into the respirator.

   (a)   General Requirements and Precautions.

      (1)   The respirator to be tested shall be equipped with high efficiency particulate air (HEPA) or P100 series filter(s).

      (2)   Only stannic chloride smoke tubes shall be used for this protocol.

      (3)   No form of test enclosure or hood for the test subject shall be used.

      (4)   The smoke can be irritating to the eyes, lungs, and nasal passages. The test conductor shall take precautions to minimize the test subject's exposure to irritant smoke. Sensitivity varies, and certain

individuals may respond to a greater degree to irritant smoke. Care shall be taken when performing the sensitivity screening checks that determine whether the test subject can detect irritant smoke to use only the minimum amount of smoke necessary to elicit a response from the test subject.

(5) The fit test shall be performed in an area with adequate ventilation to prevent exposure of the person conducting the fit test or the build-up of irritant smoke in the general atmosphere.

(b)   Sensitivity Screening Check. The person to be tested must demonstrate his or her ability to detect a weak concentration of the irritant smoke.

(1) The test operator shall break both ends of a ventilation smoke tube containing stannic chloride and attach one end of the smoke tube to a low-flow air pump set to deliver 200 milliliters per minute, or an aspirator squeeze bulb. The test operator shall cover the other end of the smoke tube with a short piece of tubing to prevent potential injury from the jagged end of the smoke tube.

(2) The test operator shall advise the test subject that the smoke can be irritating to the eyes, lungs, and nasal passages and instruct the subject to keep his or her eyes closed while the test is performed.

(3) The test subject shall be allowed to smell a weak concentration of the irritant smoke before the respirator is donned to become familiar with its irritating properties and to determine if he or she can detect the irritating properties of the smoke. The test operator shall carefully direct a small amount of the irritant smoke in the test subject's direction to determine that he or she can detect it.

(c)   Irritant Smoke Fit Test Procedure.

(1) The person being fit tested shall don the respirator without assistance, and perform the required user seal check(s).

(2) The test subject shall be instructed to keep his or her eyes closed.

(3) The test operator shall direct the stream of irritant smoke from the smoke tube toward the faceseal area of the test subject, using the low flow pump or the squeeze bulb. The test operator shall begin at least 12 inches from the facepiece and move the smoke stream around the whole perimeter of the mask. The operator shall gradually make two more passes around the perimeter of the mask, moving to within six inches of the respirator.

(4) If the person being tested has not had an involuntary response or detected the irritant smoke, proceed with the test exercises.

(5) The exercises identified in Section I.A.14. of this appendix shall be performed by the test subject while the respirator seal is being continually challenged by the smoke, directed around the perimeter of the respirator at a distance of six inches.

(6) If the person being fit tested reports detecting the irritant smoke at any time, the test is failed. The person being retested must repeat the entire sensitivity check and fit test procedure.

(7) Each test subject passing the irritant smoke test without evidence of a response (involuntary cough, irritation) shall be given a second sensitivity screening check, with the smoke from the same smoke tube used during the fit test, once the respirator has been removed, to determine whether he or she still reacts to the smoke. Failure to evoke a response shall void the fit test.

(8) If a response is produced during this second sensitivity check, then the fit test is passed.

C. Quantitative Fit Test (QNFT) protocols.

The following quantitative fit testing procedures have been demonstrated to be acceptable: Quantitative fit testing using a non-hazardous test aerosol (such as corn oil, polyethylene glycol 400 [PEG 400], di-2-ethyl hexyl sebacate [DEHS], or sodium chloride) generated in a test chamber, and employing instrumentation to quantify the fit of the respirator; quantitative fit testing using ambient aerosol as the test agent and appropriate instrumentation (condensation nuclei counter) to quantify the respirator fit; or quantitative fit testing using controlled negative pressure and appropriate instrumentation to measure the volumetric leak rate of a facepiece to quantify the respirator fit.

1. General.
   (a) The employer shall ensure that persons administering QNFT are able to calibrate equipment and perform tests properly, recognize invalid tests, calculate fit factors properly, and ensure that test equipment is in proper working order.
   (b) The employer shall ensure that QNFT equipment is kept clean, and is maintained and calibrated according to the manufacturer's instructions so as to operate at the parameters for which it was designed.

2. Generated Aerosol Quantitative Fit Testing Protocol.
   (a) Apparatus.
       (1) Instrumentation. Aerosol generation, dilution, and measurement systems using particulates (corn oil, polyethylene glycol 400 [PEG 400], di-2-ethyl hexyl sebacate [DEHS], or sodium chloride) as test aerosols shall be used for quantitative fit testing.
       (2) Test chamber. The test chamber shall be large enough to permit all test subjects to perform freely all required exercises without disturbing the test agent concentration or the measurement apparatus. The test chamber shall be equipped and constructed so that the test agent is effectively isolated from the ambient air, yet uniform in concentration throughout the chamber.
       (3) When testing air-purifying respirators, the normal filter or cartridge element shall be replaced with a high efficiency particulate air (HEPA) or P100 series filter supplied by the same manufacturer.
       (4) The sampling instrument shall be selected so that a computer record or strip chart record may be made of the test showing the rise and fall of the test agent concentration with each inspiration and

expiration at fit factors of at least 2000. Integrators or computers that integrate the amount of test agent penetration leakage into the respirator for each exercise may be used provided a record of the readings is made.

(5) The combination of substitute air-purifying elements, test agent, and test agent concentration shall be such that the test subject is not exposed in excess of an established exposure limit for the test agent at any time during the testing process, based upon the length of the exposure and the exposure limit duration.

(6) The sampling port on the test specimen respirator shall be placed and constructed so that no leakage occurs around the port (e.g., where the respirator is probed), a free air flow is allowed into the sampling line at all times, and there is no interference with the fit or performance of the respirator. The in-mask sampling device (probe) shall be designed and used so that the air sample is drawn from the breathing zone of the test subject, midway between the nose and mouth and with the probe extending into the facepiece cavity at least 1/4 inch.

(7) The test setup shall permit the person administering the test to observe the test subject inside the chamber during the test.

(8) The equipment generating the test atmosphere shall maintain the concentration of test agent constant to within a 10 percent variation for the duration of the test.

(9) The time lag (interval between an event and the recording of the event on the strip chart or computer or integrator) shall be kept to a minimum. There shall be a clear association between the occurrence of an event and its being recorded.

(10) The sampling line tubing for the test chamber atmosphere and for the respirator sampling port shall be of equal diameter and of the same material. The length of the two lines shall be equal.

(11) The exhaust flow from the test chamber shall pass through an appropriate filter (i.e., high efficiency particulate filter) before release.

(12) When sodium chloride aerosol is used, the relative humidity inside the test chamber shall not exceed 50 percent.

(13) The limitations of instrument detection shall be taken into account when determining the fit factor.

(14) Test respirators shall be maintained in proper working order and be inspected regularly for deficiencies such as cracks or missing valves and gaskets.

(b) Procedural Requirements.

(1) When performing the initial user seal check using a positive or negative pressure check, the sampling line shall be crimped closed in order to avoid air pressure leakage during either of these pressure checks.

(2) The use of an abbreviated screening QLFT test is optional. Such a test may be utilized in order to quickly identify poor fitting

respirators that passed the positive or negative pressure test and reduce the amount of QNFT time. The use of the CNC QNFT instrument in the count mode is another optional method to obtain a quick estimate of fit and eliminate poor fitting respirators before going on to perform a full QNFT.

(3) A reasonably stable test agent concentration shall be measured in the test chamber prior to testing. For canopy or shower curtain types of test units, the determination of the test agent's stability may be established after the test subject has entered the test environment.

(4) Immediately after the subject enters the test chamber, the test agent concentration inside the respirator shall be measured to ensure that the peak penetration does not exceed 5 percent for a half mask or 1 percent for a full facepiece respirator.

(5) A stable test agent concentration shall be obtained prior to the actual start of testing.

(6) Respirator restraining straps shall not be over-tightened for testing. The straps shall be adjusted by the wearer without assistance from other persons to give a reasonably comfortable fit typical of normal use. The respirator shall not be adjusted once the fit test exercises begin.

(7) The test shall be terminated whenever any single peak penetration exceeds 5 percent for half masks and 1 percent for full facepiece respirators. The test subject shall be refitted and retested.

(8) Calculation of fit factors.

   (i) The fit factor shall be determined for the quantitative fit test by taking the ratio of the average chamber concentration to the concentration measured inside the respirator for each test exercise except the grimace exercise.

   (ii) The average test chamber concentration shall be calculated as the arithmetic average of the concentration measured before and after each test (i.e., seven exercises) or the arithmetic average of the concentration measured before and after each exercise or the true average measured continuously during the respirator sample.

   (iii) The concentration of the challenge agent inside the respirator shall be determined by one of the following methods:

     (A) Average peak penetration method means the method of determining test agent penetration into the respirator utilizing a strip chart recorder, integrator, or computer. The agent penetration is determined by an average of the peak heights on the graph or by computer integration, for each exercise except the grimace exercise. Integrators or computers that calculate the actual test agent penetration into the respirator for each exercise will also be considered to meet the requirements of the average peak penetration method.

(B) Maximum peak penetration method means the method of determining test agent penetration in the respirator as determined by strip chart recordings of the test. The highest peak penetration for a given exercise is taken to be representative of average penetration into the respirator for that exercise.

(C) Integration by calculation of the area under the individual peak for each exercise except the grimace exercise. This includes computerized integration.

(D) The calculation of the overall fit factor using individual exercise fit factors involves first converting the exercise fit factors to penetration values, determining the average, and then converting that result back to a fit factor. This procedure is described in the following equation:

$$\text{Overall Fit Factor} = \frac{\text{Number of exercises}}{1/\text{ff}_1 + 1/\text{ff}_2 + 1/\text{ff}_3 + 1/\text{ff}_4 + 1/\text{ff}_5 + 1/\text{ff}_6 + 1/\text{ff}_7 + 1/\text{ff}_8}$$

Where ff_1, ff_2, ff_3, etc., are the fit factors for exercises 1, 2, 3, etc.

(9) The test subject shall not be permitted to wear a half mask or quarter facepiece respirator unless a minimum fit factor of 100 is obtained, or a full facepiece respirator unless a minimum fit factor of 500 is obtained.

(10) Filters used for quantitative fit testing shall be replaced whenever increased breathing resistance is encountered, or when the test agent has altered the integrity of the filter media.

3. Ambient aerosol condensation nuclei counter (CNC) quantitative fit testing protocol. The ambient aerosol condensation nuclei counter (CNC) quantitative fit testing (Portacount™) protocol quantitatively fit tests respirators with the use of a probe. The probed respirator is only used for quantitative fit tests. A probed respirator has a special sampling device, installed on the respirator that allows the probe to sample the air from inside the mask. A probed respirator is required for each make, style, model, and size that the employer uses and can be obtained from the respirator manufacturer or distributor. The CNC instrument manufacturer, TSI Inc., also provides probe attachments (TSI sampling adapters) that permit fit testing in an employee's own respirator. A minimum fit factor pass level of at least 100 is necessary for a half-mask respirator and a minimum fit factor pass level of at least 500 is required for a full facepiece negative pressure respirator. The entire screening and testing procedure shall be explained to the test subject prior to the conduct of the screening test.

(a) Portacount Fit Test Requirements.

(1) Check the respirator to make sure the sampling probe and line are properly attached to the facepiece and that the respirator is fitted

with a particulate filter capable of preventing significant penetration by the ambient particles used for the fit test (e.g., NIOSH 42 CFR 84 series 100, series 99, or series 95 particulate filter) per the manufacturer's instruction.

(2) Instruct the person to be tested to don the respirator for five minutes before the fit test starts. This purges the ambient particles trapped inside the respirator and permits the wearer to make certain the respirator is comfortable. This individual shall already have been trained on how to wear the respirator properly.

(3) Check the following conditions for the adequacy of the respirator fit: Chin properly placed; adequate strap tension, not overly tightened; fit across nose bridge; respirator of proper size to span distance from nose to chin; Tendency of the respirator to slip; self-observation in a mirror to evaluate fit and respirator position.

(4) Have the person wearing the respirator do a user seal check. If leakage is detected, determine the cause. If leakage is from a poorly fitting facepiece, try another size of the same model respirator, or another model of respirator.

(5) Follow the manufacturer's instructions for operating the Portacount and proceed with the test.

(6) The test subject shall be instructed to perform the exercises in Section I.A.14. of this appendix.

(7) After the test exercises, the test subject shall be questioned by the test conductor regarding the comfort of the respirator upon completion of the protocol. If it has become unacceptable, another model of respirator shall be tried.

(b) Portacount Test Instrument.

(1) The Portacount will automatically stop and calculate the overall fit factor for the entire set of exercises. The overall fit factor is what counts. The "pass" or "fail" message will indicate whether or not the test was successful. If the test was a pass, the fit test is over.

(2) Since the pass or fail criterion of the Portacount is user programmable, the test operator shall ensure that the pass or fail criterion meet the requirements for minimum respirator performance in this appendix.

(3) A record of the test needs to be kept on file, assuming the fit test was successful. The record must contain the test subject's name; overall fit factor; make, model, style, and size of respirator used; and date tested.

4. Controlled negative pressure (CNP) quantitative fit testing protocol. The CNP protocol provides an alternative to aerosol fit test methods. The CNP fit test method technology is based on exhausting air from a temporarily sealed respirator facepiece to generate and then maintain a constant negative pressure inside the facepiece. The rate of air exhaust is controlled so that a constant negative pressure is maintained in the respirator during the fit test. The level of pressure is selected to replicate the mean inspiratory pressure that causes

leakage into the respirator under normal use conditions. With pressure held constant, air flow out of the respirator is equal to air flow into the respirator. Therefore, measurement of the exhaust stream that is required to hold the pressure in the temporarily sealed respirator constant yields a direct measure of leakage air flow into the respirator. The CNP fit test method measures leak rates through the facepiece as a method for determining the facepiece fit for negative pressure respirators. The CNP instrument manufacturer Occupational Health Dynamics of Birmingham, Alabama, also provides attachments (sampling manifolds) that replace the filter cartridges to permit fit testing in an employee's own respirator. To perform the test, the test subject closes his or her mouth and holds his or her breath, after which an air pump removes air from the respirator facepiece at a preselected constant pressure. The facepiece fit is expressed as the leak rate through the facepiece, expressed as milliliters per minute. The quality and validity of the CNP fit tests are determined by the degree to which the in-mask pressure tracks the test pressure during the system measurement time of approximately five seconds. Instantaneous feedback in the form of a real-time pressure trace of the in-mask pressure is provided and used to determine test validity and quality. A minimum fit factor pass level of 100 is necessary for a half-mask respirator and a minimum fit factor of at least 500 is required for a full facepiece respirator. The entire screening and testing procedure shall be explained to the test subject prior to the conduct of the screening test.

(a)   CNP Fit Test Requirements.

    (1) The instrument shall have a non-adjustable test pressure of 15.0 millimeters of water pressure.

    (2) The CNP system defaults selected for test pressure shall be set at –15 millimeters of water (–0.58 inches of water) and the modeled inspiratory flow rate shall be 53.8 liters per minute for performing fit tests.

    (*Note*: CNP systems have built-in capability to conduct fit testing that is specific to unique work rate, mask, and gender situations that might apply in a specific workplace. Use of system default values, which were selected to represent respirator wear with medium cartridge resistance at a low-moderate work rate, will allow intertest comparison of the respirator fit.)

    (3) The individual who conducts the CNP fit testing shall be thoroughly trained to perform the test.

    (4) The respirator filter or cartridge needs to be replaced with the CNP test manifold. The inhalation valve downstream from the manifold either needs to be temporarily removed or propped open.

    (5) The employer must train the test subject to hold his or her breath for at least 10 seconds.

    (6) The test subject must don the test respirator without any assistance from the test administrator who is conducting the CNP fit test. The respirator must not be adjusted once the fit test exercises begin. Any adjustment voids the test, and the test subject must repeat the fit test.

(7) The QNFT protocol shall be followed according to Section I.C.1. of this appendix with an exception for the CNP test exercises.

(b) CNP Test Exercises.

(1) Normal breathing. In a normal standing position, without talking, the subject shall breathe normally for one minute. After the normal breathing exercise, the subject needs to hold his or her head straight ahead and hold his or her breath for 10 seconds during the test measurement.

(2) Deep breathing. In a normal standing position, the subject shall breathe slowly and deeply for one minute, being careful not to hyperventilate. After the deep breathing exercise, the subject shall hold his or her head straight ahead and hold his or her breath for 10 seconds during test measurement.

(3) Turning head side to side. Standing in place, the subject shall slowly turn his or her head from side to side between the extreme positions on each side for one minute. The head shall be held at each extreme momentarily so the subject can inhale at each side. After the turning head side to side exercise, the subject needs to hold head full left and hold his or her breath for 10 seconds during test measurement. Next, the subject needs to hold head full right and hold his or her breath for 10 seconds during test measurement.

(4) Moving head up and down. Standing in place, the subject shall slowly move his or her head up and down for one minute. The subject shall be instructed to inhale in the up position (i.e., when looking toward the ceiling). After the moving head up and down exercise, the subject shall hold his or her head full up and hold his or her breath for 10 seconds during test measurement. Next, the subject shall hold his or her head full down and hold his or her breath for 10 seconds during test measurement.

(5) Talking. The subject shall talk out loud slowly and loud enough so as to be heard clearly by the test conductor. The subject can read from a prepared text such as the Rainbow Passage, count backward from 100, or recite a memorized poem or song for one minute. After the talking exercise, the subject shall hold his or her head straight ahead and hold his or her breath for 10 seconds during the test measurement.

(6) Grimace. The test subject shall grimace by smiling or frowning for 15 seconds.

(7) Bending over. The test subject shall bend at the waist as if he or she were to touch his or her toes for one minute. Jogging in place shall be substituted for this exercise in those test environments such as shroud-type QNFT units that prohibit bending at the waist. After the bending over exercise, the subject shall hold his or her head

straight ahead and hold his or her breath for 10 seconds during the test measurement.

(8) Normal breathing. The test subject shall remove and re-don the respirator within a one-minute period. Then, in a normal standing position, without talking, the subject shall breathe normally for one minute. After the normal breathing exercise, the subject shall hold his or her head straight ahead and hold his or her breath for 10 seconds during the test measurement. After the test exercises, the test subject shall be questioned by the test conductor regarding the comfort of the respirator upon completion of the protocol. If it has become unacceptable, another model of a respirator shall be tried.

(c) CNP Test Instrument.

(1) The test instrument must have an effective audio-warning device, or a visual-warning device in the form of a screen tracing, that indicates when the test subject fails to hold his or her breath during the test. The test must be terminated and restarted from the beginning when the test subject fails to hold his or her breath during the test. The test subject then may be refitted and retested.

(2) A record of the test shall be kept on file, assuming the fit test was successful. The record must contain the test subject's name; overall fit factor; make, model, style, and size of respirator used; and date tested.

5. Controlled negative pressure (CNP) REDON quantitative fit testing protocol.

(a) When administering this protocol to test subjects, employers must comply with the requirements specified in paragraphs (a) and (c) of Part I.C.4 of this appendix ("Controlled negative pressure (CNP) quantitative fit testing protocol"), as well as use the test exercises described below in paragraph (b) of this protocol instead of the test exercises specified in paragraph (b) of Part I.C.4 of this appendix.

(b) Employers must ensure that each test subject being fit tested using this protocol follows the exercise and measurement procedures, including the order of administration, described below in Table C.1 of this appendix.

(c) After completing the test exercises, the test administrator must question each test subject regarding the comfort of the respirator. When a test subject states that the respirator is unacceptable, the employer must ensure that the test administrator repeats the protocol using another respirator model.

(d) Employers must determine the overall fit factor for each test subject by calculating the harmonic mean of the fit testing exercises as follows:

$$\text{Overall Fit Factor} = \frac{N}{[1/FF_1 + 1/FF_2 + \ldots 1/FF_N]}$$

TABLE C.1
CNP REDON Quantitative Fit Testing Protocol

Exercises*	Exercise Procedure	Measurement Procedure
Facing Forward	Stand and breathe normally, without talking, for 30 seconds.	Face forward while holding breath for 10 seconds.
Bending Over	Bend at the waist, as if going to touch his or her toes, for 30 seconds.	Face parallel to the floor while holding breath for 10 seconds
Head Shaking	For about three seconds, shake head back and forth vigorously several times while shouting.	Face forward while holding breath for 10 seconds.
Re-don 1	Remove the respirator mask, loosen all facepiece straps, and then re-don the respirator mask.	Face forward while holding breath for 10 seconds.
Re-don 2	Remove the respirator mask, loosen all facepiece straps, and then re-don the respirator mask again.	Face forward while holding breath for 10 seconds.

*Exercises are listed in the order in which they are to be administered.

Where
N   = The number of exercises
FF1 = The fit factor for the first exercise
FF2 = The fit factor for the second exercise
FFN = The fit factor for the nth exercise

PART II. NEW FIT TEST PROTOCOLS

A. Any person may submit to OSHA an application for approval of a new fit test protocol. If the application meets the following criteria, OSHA will initiate a rule-making proceeding under Section 6(b)(7) of the OSH Act to determine whether to list the new protocol as an approved protocol in this Appendix A.

B. The application must include a detailed description of the proposed new fit test protocol. This application must be supported by either:

1. A test report prepared by an independent government research laboratory (e.g., Lawrence Livermore National Laboratory, Los Alamos National Laboratory, the National Institute for Standards and Technology) stating that the laboratory has tested the protocol and had found it to be accurate and reliable

2. An article that has been published in a peer-reviewed industrial hygiene journal describing the protocol and explaining how test data support the protocol's accuracy and reliability.

C. If OSHA determines that additional information is required before the agency commences a rulemaking proceeding under this section, OSHA will so notify the applicant and afford the applicant the opportunity to submit the supplemental information. Initiation of a rulemaking proceeding will be deferred until OSHA has received and evaluated the supplemental information. [63 FR 20098, April 23, 1998; 69 FR 46993, August 4, 2004]

APPENDIX B-1 TO § 1910.134: USER SEAL CHECK PROCEDURES (MANDATORY)

The individual who uses a tight-fitting respirator is to perform a user seal check to ensure that an adequate seal is achieved each time the respirator is put on. Either the positive and negative pressure checks listed in this appendix, or the respirator manufacturer's recommended user seal check method shall be used. User seal checks are not substitutes for qualitative or quantitative fit tests.

I. FACEPIECE POSITIVE OR NEGATIVE PRESSURE CHECKS

A. *Positive pressure check.* Close off the exhalation valve and exhale gently into the facepiece. The face fit is considered satisfactory if a slight positive pressure can be built up inside the facepiece without any evidence of outward leakage of air at the seal. For most respirators this method of leak testing requires the wearer to first remove the exhalation valve cover before closing off the exhalation valve and then carefully replacing it after the test.

B. *Negative pressure check.* Close off the inlet opening of the canister or cartridge(s) by covering with the palm of the hand(s) or by replacing the filter seal(s), inhale gently so that the facepiece collapses slightly, and hold breath for ten seconds. The design of the inlet opening of some cartridges cannot be effectively covered with the palm of the hand. The test can be performed by covering the inlet opening of the cartridge with a thin latex or nitrile glove. If the facepiece remains in its slightly collapsed condition and no inward leakage of air is detected, the tightness of the respirator is considered satisfactory.

II. MANUFACTURER'S RECOMMENDED USER SEAL CHECK PROCEDURES

The respirator manufacturer's recommended procedures for performing a user seal check may be used instead of the positive or negative pressure check procedures provided that the employer demonstrates that the manufacturer's procedures are equally effective. [63 FR 1152, Jan. 8, 1998]

APPENDIX B-2 TO § 1910.134: RESPIRATOR CLEANING PROCEDURES (MANDATORY)

These procedures are provided for employer use when cleaning respirators. They are general in nature, and the employer as an alternative may use the cleaning recommendations provided by the manufacturer of the respirators used by their employees, provided such procedures are as effective as those listed here in Appendix B-2. Equivalent effectiveness simply means that the procedures used must accomplish the objectives set forth in Appendix B-2, that is, must ensure that the respirator is properly cleaned and disinfected in a manner that prevents damage to the respirator and does not cause harm to the user.

I. PROCEDURES FOR CLEANING RESPIRATORS

A. Remove filters, cartridges, or canisters. Disassemble facepieces by removing speaking diaphragms, demand and pressure-demand valve assemblies, hoses, or any components recommended by the manufacturer. Discard or repair any defective parts.

B. Wash components in warm (43°C [110°F] maximum) water with a mild detergent or with a cleaner recommended by the manufacturer. A stiff bristle (not wire) brush may be used to facilitate the removal of dirt.

C. Rinse components thoroughly in clean, warm (43°C [110°F] maximum), preferably running water. Drain.

D. When the cleaner used does not contain a disinfecting agent, respirator components should be immersed for two minutes in one of the following:

1. Hypochlorite solution (50 ppm of chlorine) made by adding approximately one milliliter of laundry bleach to one liter of water at 43°C (110°F).

2. Aqueous solution of iodine (50 ppm iodine) made by adding approximately 0.8 milliliters of tincture of iodine (six to eight grams ammonium or potassium iodide/100 cc of 45 percent alcohol) to one liter of water at 43°C (110°F).

3. Other commercially available cleansers of equivalent disinfectant quality when used as directed, if their use is recommended or approved by the respirator manufacturer.

E. Rinse components thoroughly in clean, warm (43°C [110°F] maximum), preferably running water. Drain. The importance of thorough rinsing cannot be overemphasized. Detergents or disinfectants that dry on facepieces may result in dermatitis. In addition, some disinfectants may cause deterioration of rubber or corrosion of metal parts if not completely removed.

F. Components should be hand-dried with a clean lint-free cloth or air-dried.

G. Reassemble facepiece, replacing filters, cartridges, and canisters where necessary.

H. Test the respirator to ensure that all components work properly. [63 FR 1152, Jan. 8, 1998]

APPENDIX C TO § 1910.134: OSHA RESPIRATOR MEDICAL EVALUATION QUESTIONNAIRE (MANDATORY)

To the employer: Answers to questions in Section 1 and question 9 in Section 2 of Part A do not require a medical examination.

To the employee: Your employer must allow you to answer this questionnaire during normal working hours, or at a time and place that is convenient to you. To maintain your confidentiality, your employer or supervisor must not look at or review your answers, and your employer must tell you how to deliver or send this questionnaire to the health care professional who will review it.

Part A. Section 1. (Mandatory) The following information must be provided by every employee who has been selected to use any type of respirator (please print).

1. Today's date: _____
2. Your name: _____
3. Your age (to nearest year): _____
4. Sex (circle one): Male/Female
5. Your height: _____ ft. _____ in.
6. Your weight: _____ lbs.
7. Your job title: _____
8. A phone number where you can be reached by the health care professional who reviews this questionnaire (include the area code): _____
9. The best time to phone you at this number: _____
10. Has your employer told you how to contact the health care professional who will review this questionnaire? (Circle one): Yes/No
11. Check the type of respirator you will use (you can check more than one category):
    a. _____ N, R, or P disposable respirator (filter-mask, non-cartridge type only).
    b. _____ Other type (for example, half- or full-facepiece type, powered-air purifying, supplied-air, self-contained breathing apparatus).
12. Have you worn a respirator? (Circle one): Yes/No
    If "yes," what type(s)? _____

Part A. Section 2. (Mandatory) Questions 1 through 9 below must be answered by every employee who has been selected to use any type of respirator (please circle "yes" or "no").

1. Do you currently smoke tobacco, or have you smoked tobacco in the last month: Yes/No
2. Have you ever had any of the following conditions?
   a. Seizures: Yes/No
   b. Diabetes (sugar disease): Yes/No
   c. Allergic reactions that interfere with your breathing: Yes/No
   d. Claustrophobia (fear of closed-in places): Yes/No
   e. Trouble smelling odors: Yes/No
3. Have you ever had any of the following pulmonary or lung problems?
   a. Asbestosis: Yes/No
   b. Asthma: Yes/No
   c. Chronic bronchitis: Yes/No

    d. Emphysema: Yes/No

    e. Pneumonia: Yes/No

    f. Tuberculosis: Yes/No

    g. Silicosis: Yes/No

    h. Pneumothorax (collapsed lung): Yes/No

    i. Lung cancer: Yes/No

    j. Broken ribs: Yes/No

    k. Any chest injuries or surgeries: Yes/No

    l. Any other lung problem that you've been told about: Yes/No

4. Do you currently have any of the following symptoms of pulmonary or lung illness?

    a. Shortness of breath: Yes/No

    b. Shortness of breath when walking fast on level ground or walking up a slight hill or incline: Yes/No

    c. Shortness of breath when walking with other people at an ordinary pace on level ground: Yes/No

    d. Have to stop for breath when walking at your own pace on level ground: Yes/No

    e. Shortness of breath when washing or dressing yourself: Yes/No

    f. Shortness of breath that interferes with your job: Yes/No

    g. Coughing that produces phlegm (thick sputum): Yes/No

    h. Coughing that wakes you early in the morning: Yes/No

    i. Coughing that occurs mostly when you are lying down: Yes/No

    j. Coughing up blood in the last month: Yes/No

    k. Wheezing: Yes/No

    l. Wheezing that interferes with your job: Yes/No

    m. Chest pain when you breathe deeply: Yes/No

    n. Any other symptoms that you think may be related to lung problems: Yes/No

5. Have you ever had any of the following cardiovascular or heart problems?

    a. Heart attack: Yes/No

    b. Stroke: Yes/No

    c. Angina: Yes/No

    d. Heart failure: Yes/No

    e. Swelling in your legs or feet (not caused by walking): Yes/No

    f. Heart arrhythmia (heart beating irregularly): Yes/No

    g. High blood pressure: Yes/No

    h. Any other heart problem that you've been told about: Yes/No

6. Have you ever had any of the following cardiovascular or heart symptoms?

    a. Frequent pain or tightness in your chest: Yes/No

    b. Pain or tightness in your chest during physical activity: Yes/No

    c. Pain or tightness in your chest that interferes with your job: Yes/No

    d. In the past two years, have you noticed your heart skipping or missing a beat: Yes/No

    e. Heartburn or indigestion that is not related to eating: Yes/No

f. Any other symptoms that you think may be related to heart or circulation problems: Yes/No

7. Do you currently take medication for any of the following problems?
   a. Breathing or lung problems: Yes/No
   b. Heart trouble: Yes/No
   c. Blood pressure: Yes/No
   d. Seizures: Yes/No

8. If you've used a respirator, have you ever had any of the following problems? (If you've never used a respirator, check the following space and go to question 9.)
   a. Eye irritation: Yes/No
   b. Skin allergies or rashes: Yes/No
   c. Anxiety: Yes/No
   d. General weakness or fatigue: Yes/No
   e. Any other problem that interferes with your use of a respirator: Yes/No

9. Would you like to talk to the health care professional who will review this questionnaire about your answers to this questionnaire: Yes/No

   Questions 10 to 15 below must be answered by every employee who has been selected to use either a full-facepiece respirator or a self-contained breathing apparatus (SCBA). For employees who have been selected to use other types of respirators, answering these questions is voluntary.

10. Have you ever lost vision in either eye (temporarily or permanently): Yes/No

11. Do you currently have any of the following vision problems?
    a. Wear contact lenses: Yes/No
    b. Wear glasses: Yes/No
    c. Color blind: Yes/No
    d. Any other eye or vision problem: Yes/No

12. Have you ever had an injury to your ears, including a broken ear drum: Yes/No

13. Do you currently have any of the following hearing problems?
    a. Difficulty hearing: Yes/No
    b. Wear a hearing aid: Yes/No
    c. Any other hearing or ear problem: Yes/No

14. Have you ever had a back injury: Yes/No

15. Do you currently have any of the following musculoskeletal problems?
    a. Weakness in any of your arms, hands, legs, or feet: Yes/No
    b. Back pain: Yes/No
    c. Difficulty fully moving your arms and legs: Yes/No
    d. Pain or stiffness when you lean forward or backward at the waist: Yes/No
    e. Difficulty fully moving your head up or down: Yes/No
    f. Difficulty fully moving your head side to side: Yes/No
    g. Difficulty bending at your knees: Yes/No
    h. Difficulty squatting to the ground: Yes/No
    i. Climbing a flight of stairs or a ladder carrying more than 25 pounds: Yes/No
    j. Any other muscle or skeletal problem that interferes with using a respirator: Yes/No

Part B. Any of the following questions, and other questions not listed, may be added to the questionnaire at the discretion of the health care professional who will review the questionnaire.

16. In your present job, are you working at high altitudes (over 5000 feet) or in a place that has lower than normal amounts of oxygen: Yes/No
   If "yes," do you have feelings of dizziness, shortness of breath, pounding in your chest, or other symptoms when you're working under these conditions: Yes/No

17. At work or at home, have you ever been exposed to hazardous solvents, hazardous airborne chemicals (e.g., gases, fumes, or dust), or have you come into skin contact with hazardous chemicals: Yes/No
   If "yes," name the chemicals if you know them: _____

_____

18. Have you ever worked with any of the materials, or under any of the conditions, listed below:
   a. Asbestos: Yes/No
   b. Silica (e.g., in sandblasting): Yes/No
   c. Tungsten/cobalt (e.g., grinding or welding this material): Yes/No
   d. Beryllium: Yes/No
   e. Aluminum: Yes/No
   f. Coal (e.g., mining): Yes/No
   g. Iron: Yes/No
   h. Tin: Yes/No
   i. Dusty environments: Yes/No
   j. Any other hazardous exposures: Yes/No
      If "yes," describe these exposures: _____

_____

19. List any second jobs or side businesses you have: _____

_____

20. List your previous occupations: _____

_____

21. List your current and previous hobbies: _____

_____

22. Have you been in the military services? Yes/No
   If "yes," were you exposed to biological or chemical agents (either in training or combat): Yes/No

23. Have you ever worked on a HAZMAT team? Yes/No

24. Other than medications for breathing and lung problems, heart trouble, blood pressure, and seizures mentioned earlier in this questionnaire, are you taking any other medications for any reason (including over-the-counter medications): Yes/No
   If "yes," name the medications if you know them: _____

25. Will you be using any of the following items with your respirator(s)?
    a. HEPA Filters: Yes/No
    b. Canisters (for example, gas masks): Yes/No
    c. Cartridges: Yes/No
26. How often are you expected to use the respirator(s)? (Circle "yes" or "no" for all answers that apply to you):
    a. Escape only (no rescue): Yes/No
    b. Emergency rescue only: Yes/No
    c. Less than five hours per week: Yes/No
    d. Less than two hours per day: Yes/No
    e. Two to four hours per day: Yes/No
    f. Over four hours per day: Yes/No
27. During the period you are using the respirator(s), is your work effort:
    a. Light (less than 200 kilocalories per hour): Yes/No
        If "yes," how long does this period last during the average shift:
        _____ hours _____ minutes
        Examples of a light work effort are sitting while writing, typing, drafting, or performing light assembly work; or standing while operating a drill press (one to three pounds) or controlling machines.
    b. Moderate (200 to 350 kilocalories per hour): Yes/No
        If "yes," how long does this period last during the average shift:
        _____ hours _____ minutes
        Examples of moderate work effort are sitting while nailing or filing; driving a truck or bus in urban traffic; standing while drilling, nailing, performing assembly work, or transferring a moderate load (about 35 pounds) at trunk level; walking on a level surface about two miles per hour or down a five-degree grade about three miles per hour; or pushing a wheelbarrow with a heavy load (about 100 pounds) on a level surface.
    c. Heavy (above 350 kilocalories per hour): Yes/No
        If "yes," how long does this period last during the average shift:
        _____ hours _____ minutes
        Examples of heavy work are lifting a heavy load (about 50 pounds) from the floor to your waist or shoulder; working on a loading dock; shoveling; standing while bricklaying or chipping castings; walking up an eight-degree grade about two miles per hour; climbing stairs with a heavy load (about 50 pounds).
28. Will you be wearing protective clothing and/or equipment (other than the respirator) when you're using your respirator: Yes/No
        If "yes," describe this protective clothing and/or equipment: _____

_____

29. Will you be working under hot conditions (temperature exceeding 77°F): Yes/No
30. Will you be working under humid conditions: Yes/No
31. Describe the work you'll be doing while you're using your respirator(s):

_____

_____

32. Describe any special or hazardous conditions you might encounter when you're using your respirator(s) (e.g., confined spaces, life-threatening gases):

_____

_____

33. Provide the following information, if you know it, for each toxic substance that you'll be exposed to when you're using your respirator(s):
    Name of the first toxic substance: _____
    Estimated maximum exposure level per shift: _____
    Duration of exposure per shift: _____
    Name of the second toxic substance: _____
    Estimated maximum exposure level per shift: _____
    Duration of exposure per shift: _____
    Name of the third toxic substance: _____
    Estimated maximum exposure level per shift: _____
    Duration of exposure per shift: _____
    The name of any other toxic substances that you'll be exposed to while using your respirator:

_____

_____

_____

34. Describe any special responsibilities you'll have while using your respirator(s) that may affect the safety and well-being of others (e.g., rescue, security):

_____

[63 FR 1152, Jan. 8, 1998; 63 FR 20098, April 23, 1998; 76 FR 33607, June 8, 2011; 77 FR 46949, Aug. 7, 2012]

APPENDIX D TO SEC. 1910.134 (MANDATORY) INFORMATION FOR EMPLOYEES USING RESPIRATORS WHEN NOT REQUIRED UNDER THE STANDARD

Respirators are an effective method of protection against designated hazards when properly selected and worn. Respirator use is encouraged, even when exposures are below the exposure limit, to provide an additional level of comfort and protection for workers. However, if a respirator is used improperly or not kept clean, the respirator itself can become a hazard to the worker. Sometimes, workers may wear respirators to avoid exposures to hazards, even if the amount of hazardous substance does not exceed the limits set by OSHA standards. If your employer provides respirators for your voluntary use, or if you provide your own respirator, you need to take certain precautions to be sure that the respirator itself does not present a hazard.

You should do the following:

(a) Read and heed all instructions provided by the manufacturer on use, maintenance, cleaning and care, and warnings regarding the respirator's limitations.

(b) Choose respirators certified for use to protect against the contaminant of concern. NIOSH, the National Institute for Occupational Safety and Health of the U.S. Department of Health and Human Services, certifies respirators. A label or statement of certification should appear on the respirator or respirator packaging. It will tell you what the respirator is designed for and how much it will protect you.

(c) Do not wear your respirator into atmospheres containing contaminants for which your respirator is not designed to protect against. For example, a respirator designed to filter dust particles will not protect you against gases, vapors, or very small solid particles of fumes or smoke.

(d) Keep track of your respirator so that you do not mistakenly use someone else's respirator.

[63 FR 1152, Jan. 8, 1998; 63 FR 20098, April 23, 1998]

Appendix D: Select Appendices from OSHA's Permit Required Confined Space Regulation

29 CFR § 1910.146 APPENDIX A

See Figure D.1.

29 CFR § 1910.146 APPENDIX B

Atmospheric testing is required for two distinct purposes: Evaluation of the hazards of the permit space and verification that acceptable entry conditions for entry into that space exist.

1. Evaluation testing. The atmosphere of a confined space should be analyzed using equipment of sufficient sensitivity and specificity to identify and evaluate any hazardous atmospheres that may exist or arise, so that appropriate permit entry procedures can be developed and acceptable entry conditions stipulated for that space. Evaluation and interpretation of these data, and development of the entry procedure, should be done by, or reviewed by, a technically qualified professional (e.g., OSHA consultation service, or certified industrial hygienist, registered safety engineer, certified safety professional, certified marine chemist, etc.) based on evaluation of all serious hazards.

2. Verification testing. The atmosphere of a permit space that may contain a hazardous atmosphere should be tested for residues of all contaminants identified by evaluation testing using permit-specified equipment to determine that residual concentrations at the time of testing and entry are within the range of acceptable entry conditions. Results of testing (i.e., actual concentration) should be recorded on the permit in the space provided adjacent to the stipulated acceptable entry condition.

3. Duration of testing. Measurement of values for each atmospheric parameter should be made for at least the minimum response time of the test instrument specified by the manufacturer.

4. Testing stratified atmospheres. When monitoring for entries involving a descent into atmospheres that may be stratified, the atmospheric envelope should be tested a distance of approximately four feet (1.22 meters) in the

195

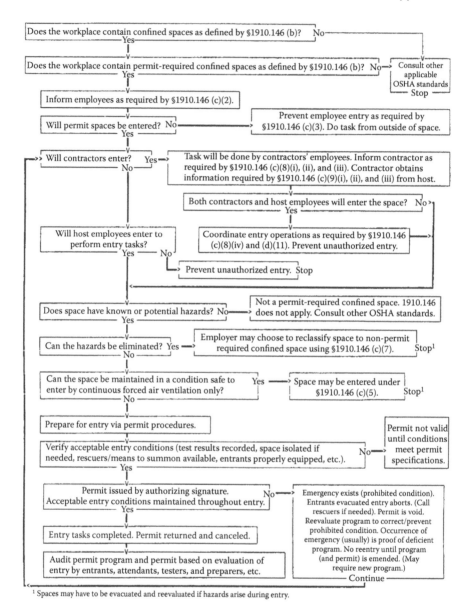

FIGURE D.1 Confined space decision flow chart.

direction of travel and to each side. If a sampling probe is used, the entrant's rate of progress should be slowed to accommodate the sampling speed and detector response.

5. Order of testing. A test for oxygen is performed first because most combustible gas meters are oxygen dependent and will not provide reliable readings in an oxygen deficient atmosphere. Combustible gases are tested for next

because the threat of fire or explosion is both more immediate and more life-threatening, in most cases, than exposure to toxic gases and vapors. If tests for toxic gases and vapors are necessary, they are performed last.

[58 FR 4549, Jan. 14, 1993; 58 FR 34846, June 29, 1993]

29 CFR § 1910.146 APPENDIX C

EXAMPLE 1

Workplace. Sewer entry.
Potential hazards. The employees could be exposed to the following:

Engulfment.
Presence of toxic gases. Equal to or more than 10 ppm hydrogen sulfide measured as an eight-hour time-weighted average. If the presence of other toxic contaminants is suspected, specific monitoring programs will be developed.
Presence of explosive/flammable gases. Equal to or greater than 10 percent of the lower flammable limit (LFL).
Oxygen deficiency. A concentration of oxygen in the atmosphere equal to or less than 19.5 percent by volume.

A. Entry without Permit/Attendant

Certification. Confined spaces may be entered without the need for a written permit or attendant provided that the space can be maintained in a safe condition for entry by mechanical ventilation alone, as provided in 1910.146(c)(5). All spaces shall be considered permit-required confined spaces until the pre-entry procedures demonstrate otherwise. Any employee required or permitted to precheck or enter an enclosed/confined space shall have successfully completed, as a minimum, the training as required by the following sections of these procedures. A written copy of operating and rescue procedures as required by these procedures shall be at the work site for the duration of the job. The Confined Space Pre-Entry Checklist must be completed by the LEAD WORKER before entry into a confined space. This list verifies completion of items listed below. This checklist shall be kept at the job site for duration of the job. If circumstances dictate an interruption in the work, the permit space must be reevaluated and a new checklist must be completed.

Control of atmospheric and engulfment hazards.

Pumps and lines. All pumps and lines which may reasonably cause contaminants to flow into the space shall be disconnected, blinded and locked out, or effectively isolated by other means to prevent development of dangerous air contamination or engulfment. Not all laterals to sewers or storm drains require blocking. However, where experience or knowledge of industrial use indicates there is a reasonable potential for contamination of air or engulfment into an occupied sewer, then all affected laterals shall be blocked. If blocking and/or isolation requires entry into the space the provisions for entry into a permit-required confined space must be implemented.

Surveillance. The surrounding area shall be surveyed to avoid hazards such as drifting vapors from the tanks, piping, or sewers.

Testing. The atmosphere within the space will be tested to determine whether dangerous air contamination or oxygen deficiency exists. Detector tubes, alarm-only gas monitors, and explosion meters are examples of monitoring equipment that may be used to test permit space atmospheres. Testing shall be performed by the LEAD WORKER who has successfully completed the Gas Detector training for the monitor he or she will use. The minimum parameters to be monitored are oxygen deficiency, LFL, and hydrogen sulfide concentration. A written record of the pre-entry test results shall be made and kept at the work site for the duration of the job. The supervisor will certify in writing, based upon the results of the pre-entry testing, that all hazards have been eliminated. Affected employees shall be able to review the testing results. The most hazardous conditions shall govern when work is being performed in two adjoining, connecting spaces.

Entry procedures. If there are no nonatmospheric hazards present and if the pre-entry tests show there is no dangerous air contamination or oxygen deficiency within the space and there is no reason to believe that any is likely to develop, entry into and work within may proceed. Continuous testing of the atmosphere in the immediate vicinity of the workers within the space shall be accomplished. The workers will immediately leave the permit space when any of the gas monitor alarm set points are reached as defined. Workers will not return to the area until a SUPERVISOR who has completed the gas detector training has used a direct reading gas detector to evaluate the situation and has determined that it is safe to enter.

Rescue. Arrangements for rescue services are not required where there is no attendant. See the rescue portion of Section B, below, for instructions regarding rescue planning where an entry permit is required.

B. Entry Permit Required

Permits. Confined Space Entry Permit. All spaces shall be considered permit-required confined spaces until the pre-entry procedures demonstrate otherwise. Any employee required or permitted to precheck or enter a permit-required confined space shall have successfully completed, as a minimum, the training as required by the following sections of these procedures. A written copy of operating and rescue procedures as required by these procedures shall be at the work site for the duration of the job. The Confined Space Entry Permit must be completed before approval can be given to enter a permit-required confined space. This permit verifies completion of items listed below. This permit shall be kept at the job site for the duration of the job. If circumstances cause an interruption in the work or a change in the alarm conditions for which entry was approved, a new Confined Space Entry Permit must be completed.

Control of atmospheric and engulfment hazards.

Surveillance. The surrounding area shall be surveyed to avoid hazards such as drifting vapors from tanks, piping, or sewers.

Testing. The confined space atmosphere shall be tested to determine whether dangerous air contamination or oxygen deficiency exists. A direct reading gas monitor shall be used. Testing shall be performed by the SUPERVISOR who has successfully completed the gas detector training for the monitor he or she will use. The minimum parameters to be monitored are oxygen deficiency, LFL, and hydrogen sulfide concentration. A written record of the pre-entry test results shall be made and kept at the work site for the duration of the job. Affected employees shall be able to review the testing results. The most hazardous conditions shall govern when work is being performed in two adjoining, connected spaces.

Space ventilation. Mechanical ventilation systems, where applicable, shall be set at 100 percent outside air. Where possible, open additional manholes to increase air circulation. Use portable blowers to augment natural circulation if needed. After a suitable ventilating period, repeat the testing. Entry may not begin until testing has demonstrated that the hazardous atmosphere has been eliminated.

Entry procedures. The following procedure shall be observed under any of the following conditions: (1) Testing demonstrates the existence of dangerous or deficient conditions and additional ventilation cannot reduce concentrations to safe levels; (2) the atmosphere tests as safe but unsafe conditions can reasonably be expected to develop; (3) it is not feasible to provide for ready exit from spaces equipped with automatic fire suppression systems and it is not practical or safe to deactivate such systems; or (4) an emergency exists and it is not feasible to wait for pre-entry procedures to take effect.

All personnel must be trained. A self-contained breathing apparatus shall be worn by any person entering the space. At least one worker shall stand by the outside of the space ready to give assistance in case of emergency. The standby worker shall have a self-contained breathing apparatus available for immediate use. There shall be at least one additional worker within sight or call of the standby worker. Continuous powered communications shall be maintained between the worker within the confined space and standby personnel.

If at any time there is any questionable action or non-movement by the worker inside, a verbal check will be made. If there is no response, the worker will be moved immediately. Exception: If the worker is disabled due to falling or impact, he or she shall not be removed from the confined space unless there is immediate danger to his or her life. Local fire department rescue personnel shall be notified immediately. The standby worker may only enter the confined space in case of an emergency (wearing the self-contained breathing apparatus) and only after being relieved by another worker. A safety belt or harness with an attached lifeline shall be used by all workers entering the space with the free end of the line secured outside the entry opening. The standby worker shall attempt to remove a disabled worker via his or her lifeline before entering the space.

When practical, these spaces shall be entered through side openings—those within 3-1/2 feet (1.07 meters) of the bottom. When entry must be through a top opening, the safety belt shall be of the harness type that suspends a person upright and a hoisting device or similar apparatus shall be available for lifting workers out of the space.

In any situation where their use may endanger the worker, use of a hoisting device or safety belt and attached lifeline may be discontinued.

When dangerous air contamination is attributable to flammable and/or explosive substances, lighting and electrical equipment shall be Class 1, Division 1 rated per National Electrical Code and no ignition sources shall be introduced into the area.

Continuous gas monitoring shall be performed during all confined space operations. If alarm conditions change adversely, entry personnel shall exit the confined space and a new confined space permit issued.

Rescue. Call the fire department services for rescue. Where immediate hazards to injured personnel are present, workers at the site shall implement emergency procedures to fit the situation.

EXAMPLE 2

Workplace. Meat- and poultry-rendering plants.

Cookers and dryers are either batch or continuous in their operation. Multiple batch cookers are operated in parallel. When one unit of a multiple set is shut down for repairs, means are available to isolate that unit from the others which remain in operation.

Cookers and dryers are horizontal, cylindrical vessels equipped with a center, rotating shaft and agitator paddles or discs. If the inner shell is jacketed, it is usually heated with steam at pressures up to 150 psig (1034.25 kPa). The rotating shaft assembly of the continuous cooker or dryer is also steam-heated.

Potential Hazards. The recognized hazards associated with cookers and dryers are the risk that employees could be

1. Struck or caught by rotating agitator.
2. Engulfed in raw material or hot, recycled fat.
3. Burned by steam from leaks into the cooker/dryer steam jacket or the condenser duct system if steam valves are not properly closed and locked out.
4. Burned by contact with hot metal surfaces, such as the agitator shaft assembly, or inner shell of the cooker/dryer.
5. Heat stress caused by warm atmosphere inside cooker/dryer.
6. Slipping and falling on grease in the cooker/dryer.
7. Electrically shocked by faulty equipment taken into the cooker/dryer.
8. Burned or overcome by fire or products of combustion.
9. Overcome by fumes generated by welding or cutting done on grease covered surfaces.

Permits. The supervisor in this case is always present at the cooker/dryer or other permit entry confined space when entry is made. The supervisor must follow the pre-entry isolation procedures described in the entry permit in preparing for entry, and ensure that the protective clothing, ventilating equipment and any other equipment required by the permit are at the entry site.

Control of hazards. Mechanical. Lock out main power switch to agitator motor at main power panel. Affix tag to the lock to inform others that a permit entry confined space entry is in progress.

Engulfment. Close all valves in the raw material blow line. Secure each valve in its closed position using chain and lock. Attach a tag to the valve and chain warning that a permit entry confined space entry is in progress. The same procedure shall be used for securing the fat recycle valve.

Burns and heat stress. Close steam supply valves to jacket and secure with chains and tags. Insert solid blank at flange in cooker vent line to condenser manifold duct system. Vent cooker/dryer by opening access door at discharge end and top center door to allow natural ventilation throughout the entry. If faster cooling is needed, use a portable ventilation fan to increase ventilation. Cooling water may be circulated through the jacket to reduce both outer and inner surface temperatures of cooker/dryers faster. Check air and inner surface temperatures in cooker/dryer to assure they are within acceptable limits before entering, or use proper protective clothing.

Fire and fume hazards. Careful site preparation, such as cleaning the area within four inches (10.16 centimeters) of all welding or torch cutting operations, and proper ventilation are the preferred controls. All welding and cutting operations shall be done in accordance with the requirements of 29 CFR Part 1910, Subpart Q, OSHA's welding standard. Proper ventilation may be achieved by local exhaust ventilation, or the use of portable ventilation fans, or a combination of the two practices.

Electrical shock. Electrical equipment used in cooker/dryers shall be in serviceable condition.

Slips and falls. Remove residual grease before entering cooker/dryer.

Attendant. The supervisor shall be the attendant for employees entering cooker/dryers.

Permit. The permit shall specify how isolation shall be done and any other preparations needed before making entry. This is especially important in parallel arrangements of cooker/dryers so that the entire operation need not be shut down to allow safe entry into one unit.

Rescue. When necessary, the attendant shall call the fire department as previously arranged.

Example 3

Workplace. Workplaces where tank cars, trucks and trailers, dry bulk tanks and trailers, railroad tank cars, and similar portable tanks are fabricated or serviced.

A. During Fabrication

These tanks and dry-bulk carriers are entered repeatedly throughout the fabrication process. These products are not configured identically, but the manufacturing processes by which they are made are very similar.

Sources of hazards. In addition to the mechanical hazards arising from the risks that an entrant would be injured due to contact with components of the tank or the tools being used, there is also the risk that a worker could be injured by breathing fumes from welding materials or mists or vapors from materials used to coat the tank interior. In addition, many of these vapors and mists are flammable, so the failure to properly ventilate a tank could lead to a fire or explosion.

Control of hazards.

Welding. Local exhaust ventilation shall be used to remove welding fumes once the tank or carrier is completed to the point that workers may enter and exit only through a manhole. (Follow the requirements of 29 CFR 1910, Subpart Q, OSHA's welding standard, at all times.) Welding gas tanks may never be brought into a tank or carrier that is a permit entry confined space.

Application of interior coatings/linings. Atmospheric hazards shall be controlled by forced air ventilation sufficient to keep the atmospheric concentration of flammable materials below 10 percent of the lower flammable limit (LFL, or lower explosive limit [LEL], whichever term is used locally). The appropriate respirators are provided and shall be used in addition to providing forced ventilation if the forced ventilation does not maintain acceptable respiratory conditions.

Permits. Because of the repetitive nature of the entries in these operations, an "Area Entry Permit" will be issued for a one-month period to cover those production areas where tanks are fabricated to the point that entry and exit are made using manholes.

Authorization. Only the area supervisor may authorize an employee to enter a tank within the permit area. The area supervisor must determine that conditions in the tank trailer, dry bulk trailer or truck, etc., meet permit requirements before authorizing entry.

Attendant. The area supervisor shall designate an employee to maintain communication by employer-specified means with employees working in tanks to ensure their safety. The attendant may not enter any permit entry confined space to rescue an entrant or for any other reason, unless authorized by the rescue procedure and, even then, only after calling the rescue team and being relieved by an attendant or another worker.

Communications and observation. Communications between attendant and entrant(s) shall be maintained throughout entry. Methods of communication that may be specified by the permit include voice, voice-powered radio, tapping or rapping codes on tank walls, signaling tugs on a rope, and the attendant's observation that work activities such as chipping, grinding, welding, spraying, etc., which require deliberate operator control continue normally. These activities often generate so much noise that the necessary hearing protection makes communication by voice difficult.

Rescue procedures. Acceptable rescue procedures include entry by a team of employee-rescuers, use of public emergency services, and procedures for breaching the tank. The area permit specifies which procedures are available, but the area supervisor makes the final decision based on circumstances. (Certain injuries may make it necessary to breach the tank to remove a person rather than risk additional injury by removal through an existing manhole. However, the supervisor must ensure that no breaching procedure used for rescue would violate terms of the entry permit. For instance, if the tank must be breached by cutting with a torch, the tank surfaces to be cut must be free of volatile or combustible coatings within four inches (10.16 centimeters) of the cutting line and the atmosphere within the tank must be below the LFL.

Retrieval line and harnesses. The retrieval lines and harnesses generally required under this standard are usually impractical for use in tanks because the internal configuration of the tanks and their interior baffles and other structures would prevent rescuers from hauling out injured entrants. However, unless the rescue procedure calls for breaching the tank for rescue, the rescue team shall be trained in the use of retrieval lines and harnesses for removing injured employees through manholes.

B. Repair or Service of "Used" Tanks and Bulk Trailers

Sources of hazards. In addition to facing the potential hazards encountered in fabrication or manufacturing, tanks or trailers which have been in service may contain residues of dangerous materials, whether left over from the transportation of hazardous cargoes or generated by chemical or bacterial action on residues of nonhazardous cargoes.

Control of atmospheric hazards. A "used" tank shall be brought into areas where tank entry is authorized only after the tank has been emptied, cleansed (without employee entry) of any residues, and purged of any potential atmospheric hazards.

Welding. In addition to tank cleaning for control of atmospheric hazards, coating and surface materials shall be removed four inches (10.16 centimeters) or more from any surface area where welding or other torch work will be done and care taken that the atmosphere within the tank remains well below the LFL. (Follow the requirements of 29 CFR 1910, Subpart Q, OSHA's welding standard, at all times.)

Permits. An entry permit valid for up to one year shall be issued prior to authorization of entry into used tank trailers, dry bulk trailers or trucks. In addition to the pre-entry cleaning requirement, this permit shall require the employee safeguards specified for new tank fabrication or construction permit areas.

Authorization. Only the area supervisor may authorize an employee to enter a tank trailer, dry bulk trailer or truck within the permit area. The area supervisor must determine that the entry permit requirements have been met before authorizing entry.

[58 FR 4549, Jan. 14, 1993; 58 FR 34846, June 29, 1993]

29 CFR § 1910.146 APPENDIX F

Non-Mandatory Appendix F—Rescue Team
or Rescue Service Evaluation Criteria

1. This appendix provides guidance to employers in choosing an appropriate rescue service. It contains criteria that may be used to evaluate the capabilities both of prospective and current rescue teams. Before a rescue team can be trained or chosen, however, a satisfactory permit program, including an analysis of all permit-required confined spaces to identify all potential hazards in those spaces, must be completed. OSHA believes that compliance with all the provisions of § 1910.146 will enable employers to conduct permit space operations without recourse to rescue services in nearly all cases. However, experience indicates that circumstances will arise where entrants will need to be rescued from permit spaces. It is therefore important for employers to select rescue services or teams, either on-site or off-site, that are equipped and capable of minimizing harm to both entrants and rescuers if the need arises.

2. For all rescue teams or services, the employer's evaluation should consist of two components: An initial evaluation, in which employers decide whether a potential rescue service or team is adequately trained and equipped to

perform permit space rescues of the kind needed at the facility and whether such rescuers can respond in a timely manner; and a performance evaluation, in which employers measure the performance of the team or service during an actual or practice rescue. For example, based on the initial evaluation, an employer may determine that maintaining an onsite rescue team will be more expensive than obtaining the services of an offsite team, without being significantly more effective, and decide to hire a rescue service. During a performance evaluation, the employer could decide, after observing the rescue service perform a practice rescue, that the service's training or preparedness was not adequate to effect a timely or effective rescue at his or her facility and decide to select another rescue service, or to form an internal rescue team.

A. Initial Evaluation

I. The employer should meet with the prospective rescue service to facilitate the evaluations required by §1910.146(k)(1)(i) and §1910.146(k)(1)(ii). At a minimum, if an offsite rescue service is being considered, the employer must contact the service to plan and coordinate the evaluations required by the standard. Merely posting the service's number or planning to rely on the 911 emergency phone number to obtain these services at the time of a permit space emergency would not comply with paragraph (k)(1) of the standard.

II. The capabilities required of a rescue service vary with the type of permit spaces from which rescue may be necessary and the hazards likely to be encountered in those spaces. Answering the questions below will assist employers in determining whether the rescue service is capable of performing rescues in the permit spaces present at the employer's workplace.

1. What are the needs of the employer with regard to response time (time for the rescue service to receive notification, arrive at the scene, and set up and be ready for entry)? For example, if entry is to be made into an IDLH atmosphere, or into a space that can quickly develop an IDLH atmosphere (if ventilation fails or for other reasons), the rescue team or service would need to be standing by at the permit space. On the other hand, if the danger to entrants is restricted to mechanical hazards that would cause injuries (e.g., broken bones, abrasions) a response time of 10 or 15 minutes might be adequate.

2. How quickly can the rescue team or service get from its location to the permit spaces from which rescue may be necessary? Relevant factors to consider would include the location of the rescue team or service relative to the employer's workplace, the quality of roads and highways to be traveled, potential bottlenecks or traffic congestion that might be encountered in transit, the reliability of the rescuer's vehicles, and the training and skill of its drivers.

3. What is the availability of the rescue service? Is it unavailable at certain times of the day or in certain situations? What is the likelihood that key personnel of the rescue service might be unavailable at times? If the rescue service becomes unavailable while an entry is underway, does it

have the capability of notifying the employer so that the employer can instruct the attendant to abort the entry immediately?

4. Does the rescue service meet all the requirements of paragraph (k)(2) of the standard? If not, has it developed a plan that will enable it to meet those requirements in the future? If so, how soon can the plan be implemented?

5. For offsite services, is the service willing to perform rescues at the employer's workplace? (An employer may not rely on a rescuer who declines, for whatever reason, to provide rescue services.)

6. Is an adequate method for communications between the attendant, employer, and prospective rescuer available so that a rescue request can be transmitted to the rescuer without delay? How soon after notification can a prospective rescuer dispatch a rescue team to the entry site?

7. For rescues into spaces that may pose significant atmospheric hazards and from which rescue entry, patient packaging, and retrieval cannot be safely accomplished in a relatively short time (15–20 minutes), employers should consider using airline respirators (with escape bottles) for the rescuers and to supply rescue air to the patient. If the employer decides to use SCBA, does the prospective rescue service have an ample supply of replacement cylinders and procedures for rescuers to enter and exit (or be retrieved) well within the SCBA's air supply limits?

8. If the space has a vertical entry more than five feet in depth, can the prospective rescue service properly perform entry rescues? Does the service have the technical knowledge and equipment to perform rope work or elevated rescue, if needed?

9. Does the rescue service have the necessary skills in medical evaluation, patient packaging and emergency response?

10. Does the rescue service have the necessary equipment to perform rescues, or must the equipment be provided by the employer or another source?

B. Performance Evaluation

Rescue services are required by paragraph (k)(2)(iv) of the standard to practice rescues at least once every 12 months, provided that the team or service has not successfully performed a permit space rescue within that time. As part of each practice session, the service should perform a critique of the practice rescue, or have another qualified party perform the critique, so that deficiencies in procedures, equipment, training, or number of personnel can be identified and corrected. The results of the critique, and the corrections made to respond to the deficiencies identified, should be given to the employer to enable it to determine whether the rescue service can quickly be upgraded to meet the employer's rescue needs or whether another service must be selected. The following questions will assist employers and rescue teams and services evaluate their performance.

1. Have all members of the service been trained as permit space entrants, at a minimum, including training in the potential hazards of all permit spaces,

or of representative permit spaces, from which rescue may be needed? Can team members recognize the signs, symptoms, and consequences of exposure to any hazardous atmospheres that may be present in those permit spaces?

2. Is every team member provided with, and properly trained in, the use and need for PPE, such as SCBA or fall-arrest equipment, which may be required to perform permit space rescues in the facility? Is every team member properly trained to perform his or her functions and make rescues, and to use any rescue equipment, such as ropes and backboards, that may be needed in a rescue attempt?

3. Are team members trained in the first aid and medical skills needed to treat victims overcome or injured by the types of hazards that may be encountered in the permit spaces at the facility?

4. Do all team members perform their functions safely and efficiently? Do rescue service personnel focus on their own safety before considering the safety of the victim?

5. If necessary, can the rescue service properly test the atmosphere to determine if it is IDLH?

6. Can the rescue personnel identify information pertinent to the rescue from entry permits, hot work permits, and MSDSs?

7. Has the rescue service been informed of any hazards to personnel that may arise from outside the space, such as those that may be caused by future work near the space?

8. If necessary, can the rescue service properly package and retrieve victims from a permit space that has a limited size opening (less than 24 inches [60.9 centimeters] in diameter), limited internal space, or internal obstacles or hazards?

9. If necessary, can the rescue service safely perform an elevated (high angle) rescue?

10. Does the rescue service have a plan for each of the kinds of permit space rescue operations at the facility? Is the plan adequate for all types of rescue operations that may be needed at the facility? Teams may practice in representative spaces, or in spaces that are "worst-case" or most restrictive with respect to internal configuration, elevation, and portal size. The following characteristics of a practice space should be considered when deciding whether a space is truly representative of an actual permit space:

    (1) Internal configuration.

        (a) Open—There are no obstacles, barriers, or obstructions within the space. One example is a water tank.

        (b) Obstructed—The permit space contains some type of obstruction that a rescuer would need to maneuver around. An example would be a baffle or mixing blade. Large equipment, such as a ladder or scaffold, brought into a space for work purposes would be considered an obstruction if the positioning or size of the equipment would make rescue more difficult.

(2) Elevation.
  (a) Elevated—A permit space where the entrance portal or opening is above grade by four feet or more. This type of space usually requires knowledge of high angle rescue procedures because of the difficulty in packaging and transporting a patient to the ground from the portal.
  (b) Non-elevated—A permit space with the entrance portal located less than four feet above grade. This type of space will allow the rescue team to transport an injured employee normally.
(3) Portal size.
  (a) Restricted—A portal of 24 inches or less in the least dimension. Portals of this size are too small to allow a rescuer to simply enter the space while using SCBA. The portal size is also too small to allow normal spinal immobilization of an injured employee.
  (b) Unrestricted—A portal of greater than 24 inches in the least dimension. These portals allow relatively free movement into and out of the permit space.
(4) Space access.
  (a) Horizontal—The portal is located on the side of the permit space. Use of retrieval lines could be difficult.
  (b) Vertical—The portal is located on the top of the permit space, so that rescuers must climb down, or the bottom of the permit space, so that rescuers must climb up to enter the space. Vertical portals may require knowledge of rope techniques, or special patient packaging to safely retrieve a downed entrant.

[63 FR 66039, Dec. 1, 1998]

Appendix E: OSHA's New Rules on Respirable Silica for General Industry and Construction

GENERAL INDUSTRY/MARITIME

1910.1053(a)
Scope and application

1910.1053(a)(1)
This section applies to all occupational exposures to respirable crystalline silica, except:

1910.1053(a)(1)(i)
Construction work as defined in 29 CFR 1910.12(b) (occupational exposures to respirable crystalline silica in construction work are covered under 29 CFR 1926.1153).

1910.1053(a)(1)(ii)
Agricultural operations covered under 29 CFR part 1928.

1910.1053(a)(1)(iii)
Exposures that result from the processing of sorptive clays.

1910.1053(a)(2)
This section does not apply where the employer has objective data demonstrating that employee exposure to respirable crystalline silica will remain below 25 micrograms per cubic meter of air (25 µg/m^3) as an eight-hour time-weighted average (TWA) under any foreseeable conditions.

1910.1053(a)(3)
This section does not apply if the employer complies with 29 CFR 1926.1153 and

1910.1053(a)(3)(i)
The task performed is indistinguishable from a construction task listed on Table E.1 in paragraph (c) of 29 CFR 1926.1153, and

1910.1053(a)(3)(ii)
The task will not be performed regularly in the same environment and conditions.

1910.1053(b)

Definitions. For the purposes of this section, the following definitions apply.

Action level means a concentration of airborne respirable crystalline silica of 25 µg/m³, calculated as an eight-hour TWA.

Assistant Secretary means the Assistant Secretary of Labor for Occupational Safety and Health, U.S. Department of Labor, or designee.

Director means the Director of the National Institute for Occupational Safety and Health (NIOSH), U.S. Department of Health and Human Services, or designee.

Employee exposure means the exposure to airborne respirable crystalline silica that would occur if the employee were not using a respirator.

High-efficiency particulate air [HEPA] filter means a filter that is at least 99.97 percent efficient in removing monodispersed particles of 0.3 micrometers in diameter.

Objective data means information, such as air monitoring data from industry-wide surveys or calculations based on the composition of a substance, demonstrating employee exposure to respirable crystalline silica associated with a particular product or material or a specific process, task, or activity. The data must reflect workplace conditions closely resembling or with a higher exposure potential than the processes, types of material, control methods, work practices, and environmental conditions in the employer's current operations.

Physician or other licensed health care professional [PLHCP] means an individual whose legally permitted scope of practice (i.e., license, registration, or certification) allows him or her to independently provide or be delegated the responsibility to provide some or all of the particular health care services required by paragraph (i) of this section.

Regulated area means an area, demarcated by the employer, where an employee's exposure to airborne concentrations of respirable crystalline silica exceeds, or can reasonably be expected to exceed, the PEL.

Respirable crystalline silica means quartz, cristobalite, or tridymite contained in airborne particles that are determined to be respirable by a sampling device designed to meet the characteristics for respirable-particle size-selective samplers specified in the International Organization for Standardization (ISO) 7708:1995: Air Quality-Particle Size Fraction Definitions for Health-Related Sampling.

Specialist means an American Board Certified Specialist in Pulmonary Disease or an American Board Certified Specialist in Occupational Medicine.

This section means this respirable crystalline silica standard, 29 CFR 1910.1053.

1910.1053(c)

Permissible exposure limit (PEL). The employer shall ensure that no employee is exposed to an airborne concentration of respirable crystalline silica in excess of 50 µg/m³, calculated as an eight-hour TWA.

1910.1053(d)

Exposure assessment.

1910.1053(d)(1)

General. The employer shall assess the exposure of each employee who is or may reasonably be expected to be exposed to respirable crystalline silica at or above the action level in accordance with either the performance option in paragraph (d)(2) or the scheduled monitoring option in paragraph (d)(3) of this section.

1910.1053(d)(2)

Performance option. The employer shall assess the eight-hour TWA exposure for each employee on the basis of any combination of air monitoring data or objective data sufficient to accurately characterize employee exposures to respirable crystalline silica.

1910.1053(d)(3)

Scheduled monitoring option.

1910.1053(d)(3)(i)

The employer shall perform initial monitoring to assess the eight-hour TWA exposure for each employee on the basis of one or more personal breathing zone air samples that reflect the exposures of employees on each shift, for each job classification, in each work area. Where several employees perform the same tasks on the same shift and in the same work area, the employer may sample a representative fraction of these employees in order to meet this requirement. In representative sampling, the employer shall sample the employee(s) who are expected to have the highest exposure to respirable crystalline silica.

1910.1053(d)(3)(ii)

If initial monitoring indicates that employee exposures are below the action level, the employer may discontinue monitoring for those employees whose exposures are represented by such monitoring.

1910.1053(d)(3)(iii)

Where the most recent exposure monitoring indicates that employee exposures are at or above the action level but at or below the PEL, the employer shall repeat such monitoring within six months of the most recent monitoring.

1910.1053(d)(3)(iv)

Where the most recent exposure monitoring indicates that employee exposures are above the PEL, the employer shall repeat such monitoring within three months of the most recent monitoring.

1910.1053(d)(3)(v)

Where the most recent (non-initial) exposure monitoring indicates that employee exposures are below the action level, the employer shall repeat such monitoring within six months of the most recent monitoring until two consecutive measurements, taken seven or more days apart, are below the action level, at which time the employer may discontinue monitoring for those employees whose exposures are represented by such monitoring, except as otherwise provided in paragraph (d)(4) of this section.

1910.1053(d)(4)

Reassessment of exposures. The employer shall reassess exposures whenever a change in the production, process, control equipment, personnel, or work practices may reasonably be expected to result in new or additional exposures at or above the action level, or when the employer has any reason to believe that new or additional exposures at or above the action level have occurred.

1910.1053(d)(5)

Methods of sample analysis. The employer shall ensure that all samples taken to satisfy the monitoring requirements of paragraph (d) of this section are evaluated by a laboratory that analyzes air samples for respirable crystalline silica in accordance with the procedures in Appendix A to this section.

1910.1053(d)(6)

Employee notification of assessment results.

1910.1053(d)(6)(i)

Within 15 working days after completing an exposure assessment in accordance with paragraph (d) of this section, the employer shall individually notify each affected employee in writing of the results of that assessment or post the results in an appropriate location accessible to all affected employees.

1910.1053(d)(6)(ii)

Whenever an exposure assessment indicates that employee exposure is above the PEL, the employer shall describe in the written notification the corrective action being taken to reduce employee exposure to or below the PEL.

1910.1053(d)(7)

Observation of monitoring.

1910.1053(d)(7)(i)

Where air monitoring is performed to comply with the requirements of this section, the employer shall provide affected employees or their designated representatives an opportunity to observe any monitoring of employee exposure to respirable crystalline silica.

1910.1053(d)(7)(ii)

When observation of monitoring requires entry into an area where the use of protective clothing or equipment is required for any workplace hazard, the employer shall provide the observer with protective clothing and equipment at no cost and shall ensure that the observer uses such clothing and equipment.

1910.1053(e)

Regulated areas.

1910.1053(e)(1)

Establishment. The employer shall establish a regulated area wherever an employee's exposure to airborne concentrations of respirable crystalline silica is, or can reasonably be expected to be, in excess of the PEL.

1910.1053(e)(2)

Demarcation.

1910.1053(e)(2)(i)

The employer shall demarcate regulated areas from the rest of the workplace in a manner that minimizes the number of employees exposed to respirable crystalline silica within the regulated area.

1910.1053(e)(2)(ii)

The employer shall post signs at all entrances to regulated areas that bear the legend specified in paragraph (j)(2) of this section.

1910.1053(e)(3)

Access. The employer shall limit access to regulated areas to

1910.1053(e)(3)(A)

Persons authorized by the employer and required by work duties to be present in the regulated area.

1910.1053(e)(3)(B)

Any person entering such an area as a designated representative of employees for the purpose of exercising the right to observe monitoring procedures under paragraph (d) of this section.

1910.1053(e)(3)(C)

Any person authorized by the Occupational Safety and Health Act or regulations issued under it to be in a regulated area.

1910.1053(e)(4)

Provision of respirators. The employer shall provide each employee and the employee's designated representative entering a regulated area with an appropriate respirator in accordance with paragraph (g) of this section and shall require each employee and the employee's designated representative to use the respirator while in a regulated area.

1910.1053(f)

Methods of compliance.

1910.1053(f)(1)

Engineering and work practice controls. The employer shall use engineering and work practice controls to reduce and maintain employee exposure to respirable crystalline silica to or below the PEL, unless the employer can demonstrate that such controls are not feasible. Wherever such feasible engineering and work practice controls are not sufficient to reduce employee exposure to or below the PEL, the employer shall nonetheless use them to reduce employee exposure to the lowest feasible level and shall supplement them with the use of respiratory protection that complies with the requirements of paragraph (g) of this section.

1910.1053(f)(2)

Written exposure control plan.

1910.1053(f)(2)(i)

The employer shall establish and implement a written exposure control plan that contains at least the following elements:

1910.1053(f)(2)(i)(A)

A description of the tasks in the workplace that involve exposure to respirable crystalline silica.

1910.1053(f)(2)(i)(B)
A description of the engineering controls, work practices, and respiratory protection used to limit employee exposure to respirable crystalline silica for each task.

1910.1053(f)(2)(i)(C)
A description of the housekeeping measures used to limit employee exposure to respirable crystalline silica.

1910.1053(f)(2)(ii)
The employer shall review and evaluate the effectiveness of the written exposure control plan at least annually and update it as necessary.

1910.1053(f)(2)(iii)
The employer shall make the written exposure control plan readily available for examination and copying, upon request, to each employee covered by this section, their designated representatives, the Assistant Secretary and the Director.

1910.1053(f)(3)
Abrasive blasting. In addition to the requirements of paragraph (f)(1) of this section, the employer shall comply with other OSHA standards, when applicable, such as 29 CFR 1910.94 (Ventilation), 29 CFR 1915.34 (Mechanical paint removers), and 29 CFR 1915 Subpart I (Personal Protective Equipment), where abrasive blasting is conducted using crystalline silica–containing blasting agents, or where abrasive blasting is conducted on substrates that contain crystalline silica.

1910.1053(g)
Respiratory protection.

1910.1053(g)(1)
General. Where respiratory protection is required by this section, the employer must provide each employee an appropriate respirator that complies with the requirements of this paragraph and 29 CFR 1910.134. Respiratory protection is required:

1910.1053(g)(1)(i)
Where exposures exceed the PEL during periods necessary to install or implement feasible engineering and work practice controls.

1910.1053(g)(1)(ii)
Where exposures exceed the PEL during tasks, such as certain maintenance and repair tasks, for which engineering and work practice controls are not feasible.

1910.1053(g)(1)(iii)
During tasks for which an employer has implemented all feasible engineering and work practice controls and such controls are not sufficient to reduce exposures to or below the PEL.

1910.1053(g)(1)(iv)
During periods when the employee is in a regulated area.

1910.1053(g)(2)
Respiratory protection program. Where respirator use is required by this section, the employer shall institute a respiratory protection program in accordance with 29 CFR 1910.134.

1910.1053(h)
Housekeeping.

1910.1053(h)(1)
The employer shall not allow dry sweeping or dry brushing where such activity could contribute to employee exposure to respirable crystalline silica unless wet sweeping, HEPA-filtered vacuuming or other methods that minimize the likelihood of exposure are not feasible.

1910.1053(h)(2)
The employer shall not allow compressed air to be used to clean clothing or surfaces where such activity could contribute to employee exposure to respirable crystalline silica unless:

1910.1053(h)(2)(i)
The compressed air is used in conjunction with a ventilation system that effectively captures the dust cloud created by the compressed air; or

1910.1053(h)(2)(ii)
No alternative method is feasible.

1910.1053(i)
Medical surveillance.

1910.1053(i)(1)
General.

1910.1053(i)(1)(i)
The employer shall make medical surveillance available at no cost to the employee, and at a reasonable time and place, for each employee who will be occupationally exposed to respirable crystalline silica at or above the action level for 30 or more days per year.

1910.1053(i)(1)(ii)
The employer shall ensure that all medical examinations and procedures required by this section are performed by a PLHCP as defined in paragraph (b) of this section.

1910.1053(i)(2)
Initial examination. The employer shall make available an initial (baseline) medical examination within 30 days after initial assignment, unless the employee has received a medical examination that meets the requirements of this section within the last three years. The examination shall consist of

1910.1053(i)(2)(i)

A medical and work history, with emphasis on past, present, and anticipated exposure to respirable crystalline silica, dust, and other agents affecting the respiratory system; any history of respiratory system dysfunction, including signs and symptoms of respiratory disease (e.g., shortness of breath, coughing, wheezing); history of tuberculosis; and smoking status and history.

1910.1053(i)(2)(ii)

A physical examination with special emphasis on the respiratory system.

1910.1053(i)(2)(iii)

A chest x-ray (a single posteroanterior radiographic projection or radiograph of the chest at full inspiration recorded on either film (no less than 14 × 17 inches and no more than 16 × 17 inches) or digital radiography systems), interpreted and classified according to the International Labor Office (ILO) International Classification of Radiographs of Pneumoconiosis by a NIOSH-certified B Reader.

1910.1053(i)(2)(iv)

A pulmonary function test to include forced vital capacity (FVC) and forced expiratory volume in one second (FEV1) and FEV1/FVC ratio, administered by a spirometry technician with a current certificate from a NIOSHapproved spirometry course.

1910.1053(i)(2)(v)

Testing for latent tuberculosis infection.

1910.1053(i)(2)(vi)

Any other tests deemed appropriate by the PLHCP.

1910.1053(i)(3)

Periodic examinations. The employer shall make available medical examinations that include the procedures described in paragraph (i)(2) of this section (except paragraph (i)(2)(v)) at least every three years, or more frequently if recommended by the PLHCP.

1910.1053(i)(4)

Information provided to the PLHCP. The employer shall ensure that the examining PLHCP has a copy of this standard, and shall provide the PLHCP with the following information:

1910.1053(i)(4)(i)

A description of the employee's former, current, and anticipated duties as they relate to the employee's occupational exposure to respirable crystalline silica.

1910.1053(i)(4)(ii)

The employee's former, current, and anticipated levels of occupational exposure to respirable crystalline silica.

1910.1053(i)(4)(iii)

A description of any personal protective equipment used or to be used by the employee, including when and for how long the employee has used or will use that equipment.

1910.1053(i)(4)(iv)
Information from records of employment-related medical examinations previously provided to the employee and currently within the control of the employer.

1910.1053(i)(5)
PLHCP's written medical report for the employee. The employer shall ensure that the PLHCP explains to the employee the results of the medical examination and provides each employee with a written medical report within 30 days of each medical examination performed. The written report shall contain:

1910.1053(i)(5)(i)
A statement indicating the results of the medical examination, including any medical condition(s) that would place the employee at increased risk of material impairment to health from exposure to respirable crystalline silica and any medical conditions that require further evaluation or treatment.

1910.1053(i)(5)(ii)
Any recommended limitations on the employee's use of respirators.

1910.1053(i)(5)(iii)
Any recommended limitations on the employee's exposure to respirable crystalline silica.

1910.1053(i)(5)(iv)
A statement that the employee should be examined by a specialist (pursuant to paragraph (i)(7) of this section) if the chest x-ray provided in accordance with this section is classified as 1/0 or higher by the B Reader, or if referral to a specialist is otherwise deemed appropriate by the PLHCP.

1910.1053(i)(6)
PLHCP's written medical opinion for the employer.

1910.1053(i)(6)(i)
The employer shall obtain a written medical opinion from the PLHCP within 30 days of the medical examination. The written opinion shall contain only the following:

1910.1053(i)(6)(i)(A)
The date of the examination.

1910.1053(i)(6)(i)(B)
A statement that the examination has met the requirements of this section.

1910.1053(i)(6)(i)(C)
Any recommended limitations on the employee's use of respirators.

1910.1053(i)(6)(ii)
If the employee provides written authorization, the written opinion shall also contain either or both of the following:

1910.1053(i)(6)(ii)(A)
Any recommended limitations on the employee's exposure to respirable crystalline silica.

1910.1053(i)(6)(ii)(B)

A statement that the employee should be examined by a specialist (pursuant to paragraph (i)(7) of this section) if the chest x-ray provided in accordance with this section is classified as 1/0 or higher by the B Reader, or if referral to a specialist is otherwise deemed appropriate by the PLHCP.

1910.1053(i)(6)(iii)

The employer shall ensure that each employee receives a copy of the written medical opinion described in paragraph (i)(6)(i) and (ii) of this section within 30 days of each medical examination performed.

1910.1053(i)(7)

Additional examinations.

1910.1053(i)(7)(i)

If the PLHCP's written medical opinion indicates that an employee should be examined by a specialist, the employer shall make available a medical examination by a specialist within 30 days after receiving the PLHCP's written opinion.

1910.1053(i)(7)(ii)

The employer shall ensure that the examining specialist is provided with all of the information that the employer is obligated to provide to the PLHCP in accordance with paragraph (i)(4) of this section.

1910.1053(i)(7)(iii)

The employer shall ensure that the specialist explains to the employee the results of the medical examination and provides each employee with a written medical report within 30 days of the examination. The written report shall meet the requirements of paragraph (i)(5) (except paragraph (i)(5)(iv)) of this section.

1910.1053(i)(7)(iv)

The employer shall obtain a written opinion from the specialist within 30 days of the medical examination. The written opinion shall meet the requirements of paragraph (i)(6) (except paragraph (i)(6)(i)(B) and (i)(6)(ii)(B)) of this section.

1910.1053(j)

Communication of respirable crystalline silica hazards to employees.

1910.1053(j)(1)

Hazard communication. The employer shall include respirable crystalline silica in the program established to comply with the hazard communication standard (HCS) (29 CFR 1910.1200). The employer shall ensure that each employee has access to labels on containers of crystalline silica and safety data sheets, and is trained in accordance with the provisions of HCS and paragraph (j)(3) of this section. The employer shall ensure that at least the following hazards are addressed: Cancer, lung effects, immune system effects, and kidney effects.

1910.1053(j)(2)

Signs. The employer shall post signs at all entrances to regulated areas that bear the following legend:

DANGER
RESPIRABLE CRYSTALLINE SILICA
MAY CAUSE CANCER
CAUSES DAMAGE TO LUNGS
WEAR RESPIRATORY PROTECTION IN
THIS AREA
AUTHORIZED PERSONNEL ONLY

1910.1053(j)(3)

Employee information and training.

1910.1053(j)(3)(i)

The employer shall ensure that each employee covered by this section can demonstrate knowledge and understanding of at least the following:

1910.1053(j)(3)(i)(A)

The health hazards associated with exposure to respirable crystalline silica.

1910.1053(j)(3)(i)(B)

Specific tasks in the workplace that could result in exposure to respirable crystalline silica.

1910.1053(j)(3)(i)(C)

Specific measures the employer has implemented to protect employees from exposure to respirable crystalline silica, including engineering controls, work practices, and respirators to be used.

1910.1053(j)(3)(i)(D)

The contents of this section.

1910.1053(j)(3)(i)(E)

The purpose and a description of the medical surveillance program required by paragraph (i) of this section.

1910.1053(j)(3)(ii)

The employer shall make a copy of this section readily available without cost to each employee covered by this section.

1910.1053(k)

Recordkeeping.

1910.1053(k)(1)

Air monitoring data.

1910.1053(k)(1)(i)
The employer shall make and maintain an accurate record of all exposure measurements taken to assess employee exposure to respirable crystalline silica, as prescribed in paragraph (d) of this section.

1910.1053(k)(1)(ii)
This record shall include at least the following information:

1910.1053(k)(1)(ii)(A)
The date of measurement for each sample taken.

1910.1053(k)(1)(ii)(B)
The task monitored.

1910.1053(k)(1)(ii)(C)
Sampling and analytical methods used.

1910.1053(k)(1)(ii)(D)
Number, duration, and results of samples taken.

1910.1053(k)(1)(ii)(E)
Identity of the laboratory that performed the analysis.

1910.1053(k)(1)(ii)(F)
Type of personal protective equipment, such as respirators, worn by the employees monitored.

1910.1053(k)(1)(ii)(G)
Name, social security number, and job classification of all employees represented by the monitoring, indicating which employees were actually monitored.

1910.1053(k)(1)(iii)
The employer shall ensure that exposure records are maintained and made available in accordance with 29 CFR 1910.1020.

1910.1053(k)(2)
Objective data.

1910.1053(k)(2)(i)
The employer shall make and maintain an accurate record of all objective data relied upon to comply with the requirements of this section.

1910.1053(k)(2)(ii)
This record shall include at least the following information:

1910.1053(k)(2)(ii)(A)
The crystalline silica-containing material in question.

1910.1053(k)(2)(ii)(B)
The source of the objective data.

1910.1053(k)(2)(ii)(C)
The testing protocol and results of testing.

1910.1053(k)(2)(ii)(D)
A description of the process, task, or activity on which the objective data were based.

1910.1053(k)(2)(ii)(E)
Other data relevant to the process, task, activity, material, or exposures on which the objective data were based.

1910.1053(k)(2)(iii)
The employer shall ensure that objective data are maintained and made available in accordance with 29 CFR 1910.1020.

1910.1053(k)(3)
Medical surveillance.

1910.1053(k)(3)(i)
The employer shall make and maintain an accurate record for each employee covered by medical surveillance under paragraph (i) of this section.

1910.1053(k)(3)(ii)
The record shall include the following information about the employee:

1910.1053(k)(3)(ii)(A)
Name and social security number.

1910.1053(k)(3)(ii)(B)
A copy of the PLHCPs' and specialists' written medical opinions.

1910.1053(k)(3)(ii)(C)
A copy of the information provided to the PLHCPs and specialists.

1910.1053(k)(3)(iii)
The employer shall ensure that medical records are maintained and made available in accordance with 29 CFR 1910.1020.

1910.1053(l)
Dates.

1910.1053(l)(1)
This section is effective June 23, 2016.

1910.1053(l)(2)
Except as provided for in paragraphs (l)(3) and (4) of this section, all obligations of this section commence June 23, 2018.

1910.1053(l)(3)
For hydraulic fracturing operations in the oil and gas industry:

1910.1053(l)(3)(i)

All obligations of this section, except obligations for medical surveillance in paragraph (i)(1)(i) and engineering controls in paragraph (f)(1) of this section, commence June 23, 2018.

1910.1053(l)(3)(ii)

Obligations for engineering controls in paragraph (f)(1) of this section commence June 23, 2021.

1910.1053(l)(3)(iii)

Obligations for medical surveillance in paragraph (i)(1)(i) commence in accordance with paragraph (l)(4) of this section.

1910.1053(l)(4)

The medical surveillance obligations in paragraph (i)(1)(i) commence on June 23, 2018, for employees who will be occupationally exposed to respirable crystalline silica above the PEL for 30 or more days per year. Those obligations commence June 23, 2020, for employees who will be occupationally exposed to respirable crystalline silica at or above the action level for 30 or more days per year.

CONSTRUCTION

1926.1153(a)

Scope and application. This section applies to all occupational exposures to respirable crystalline silica in construction work, except where employee exposure will remain below 25 micrograms per cubic meter of air (25 µg/m³) as an eight-hour time-weighted average (TWA) under any foreseeable conditions.

1926.1153(b)

Definitions. For the purposes of this section the following definitions apply:

Action level means a concentration of airborne respirable crystalline silica of 25 µg/m³, calculated as an eight-hour TWA.

Assistant Secretary means the Assistant Secretary of Labor for Occupational Safety and Health, U.S. Department of Labor, or designee.

Director means the Director of the National Institute for Occupational Safety and Health (NIOSH), U.S. Department of Health and Human Services, or designee.

Competent person means an individual who is capable of identifying existing and foreseeable respirable crystalline silica hazards in the workplace and who has authorization to take prompt corrective measures to eliminate or minimize them. The competent person must have the knowledge and ability necessary to fulfill the responsibilities set forth in paragraph (g) of this section.

Employee exposure means the exposure to airborne respirable crystalline silica that would occur if the employee were not using a respirator.

High-efficiency particulate air [HEPA] filter means a filter that is at least 99.97 percent efficient in removing monodispersed particles of 0.3 micrometers in diameter.

Objective data means information, such as air monitoring data from industry-wide surveys or calculations based on the composition of a substance, demonstrating employee exposure to respirable crystalline silica associated with a particular product or material or a specific process, task, or activity. The data must reflect workplace conditions closely resembling or with a higher exposure potential than the processes, types of material, control methods, work practices, and environmental conditions in the employer's current operations.

Physician or other licensed health care professional [PLHCP] means an individual whose legally permitted scope of practice (i.e., license, registration, or certification) allows him or her to independently provide or be delegated the responsibility to provide some or all of the particular health care services required by paragraph (h) of this section.

Respirable crystalline silica means quartz, cristobalite, or tridymite contained in airborne particles that are determined to be respirable by a sampling device designed to meet the characteristics for respirable-particle size-selective samplers specified in the International Organization for Standardization (ISO) 7708:1995: Air Quality–Particle Size Fraction Definitions for Health-Related Sampling.

Specialist means an American Board Certified Specialist in Pulmonary Disease or an American Board Certified Specialist in Occupational Medicine.

This section means this respirable crystalline silica standard, 29 CFR 1926.1153.

1926.1153(c)
Specified exposure control methods.

1926.1153(c)(1)
For each employee engaged in a task identified on Table E.1, the employer shall fully and properly implement the engineering controls, work practices, and respiratory protection specified for the task on Table E.1, unless the employer assesses and limits the exposure of the employee to respirable crystalline silica in accordance with paragraph (d) of this section.

1926.1153(c)(2)
When implementing the control measures specified in Table E.1, each employer shall

1926.1153(c)(2)(i)
For tasks performed indoors or in enclosed areas, provide a means of exhaust as needed to minimize the accumulation of visible airborne dust.

1926.1153(c)(2)(ii)
For tasks performed using wet methods, apply water at flow rates sufficient to minimize release of visible dust.

1926.1153(c)(2)(iii)
For measures implemented that include an enclosed cab or booth, ensure that the enclosed cab or booth:

1926.1153(c)(2)(iii)(A)
Is maintained as free as practicable from settled dust.

1926.1153(c)(2)(iii)(B)
Has door seals and closing mechanisms that work properly.

TABLE E.1
Specified Exposure Control Methods When Working with Materials Containing Crystalline Silica

Equipment/Task	Engineering and Work Practice Control Methods	Required Respiratory Protection and Minimum Assigned Protection Factor (APF)	
		≤4 Hours/ Shift	>4 Hours/ Shift
(i) Stationary masonry saws	Use saw equipped with integrated water delivery system that continuously feeds water to the blade.	None	None
	Operate and maintain tool in accordance with manufacturer's instructions to minimize dust emissions.		
(ii) Handheld power saws (any blade diameter)	Use saw equipped with integrated water delivery system that continuously feeds water to the blade.		
	Operate and maintain tool in accordance with manufacturer's instructions to minimize dust emissions.		
	– When used outdoors.	None	APF 10
	– When used indoors or in an enclosed area.	APF 10	APF 10
(iii) Handheld power saws for cutting fiber-cement board (with blade diameter of eight inches or less)	For tasks performed outdoors only:		
	Use saw equipped with commercially available dust collection system.	None	None
	Operate and maintain tool in accordance with manufacturer's instructions to minimize dust emissions.		
	Dust collector must provide the air flow recommended by the tool manufacturer, or greater, and have a filter with 99 percent or greater efficiency.		
(iv) Walk-behind saws	Use saw equipped with integrated water delivery system that continuously feeds water to the blade.		
	Operate and maintain tool in accordance with manufacturer's instructions to minimize dust emissions.		
	– When used outdoors.	None	None
	– When used indoors or in an enclosed area.	APF 10	APF 10

(*Continued*)

TABLE E.1 (CONTINUED)
Specified Exposure Control Methods When Working with Materials
Containing Crystalline Silica

Equipment/Task	Engineering and Work Practice Control Methods	Required Respiratory Protection and Minimum Assigned Protection Factor (APF)	
		≤4 Hours/ Shift	>4 Hours/ Shift
(v) Drivable saws	For tasks performed outdoors only:		
	Use saw equipped with integrated water delivery system that continuously feeds water to the blade.	None	None
	Operate and maintain tool in accordance with manufacturer's instructions to minimize dust emissions.		
(vi) Rig-mounted core saws or drills	Use tool equipped with integrated water delivery system that supplies water to cutting surface.	None	None
	Operate and maintain tool in accordance with manufacturer's instructions to minimize dust emissions.		
(vii) Handheld and stand-mounted drills (including impact and rotary hammer drills)	Use drill equipped with commercially available shroud or cowling with dust collection system.	None	None
	Operate and maintain tool in accordance with manufacturer's instructions to minimize dust emissions.		
	Dust collector must provide the air flow recommended by the tool manufacturer, or greater, and have a filter with 99 percent or greater efficiency and a filter-cleaning mechanism.		
	Use a HEPA-filtered vacuum when cleaning holes.		
(viii) Dowel drilling rigs for concrete	For tasks performed outdoors only:		
	Use shroud around drill bit with a dust collection system. Dust collector must have a filter with 99 percent or greater efficiency and a filter-cleaning mechanism.	APF 10	APF 10
	Use a HEPA-filtered vacuum when cleaning holes.		

(Continued)

TABLE E.1 (CONTINUED)

Specified Exposure Control Methods When Working with Materials Containing Crystalline Silica

Equipment/Task	Engineering and Work Practice Control Methods	Required Respiratory Protection and Minimum Assigned Protection Factor (APF)	
		≤4 Hours/ Shift	>4 Hours/ Shift
(ix) Vehicle-mounted drilling rigs for rock and concrete	Use dust collection system with close capture hood or shroud around drill bit with a low-flow water spray to wet the dust at the discharge point from the dust collector. OR	None	None
	Operate from within an enclosed cab and use water for dust suppression on drill bit.	None	None
(x) Jackhammers and handheld powered chipping tools	Use tool with water delivery system that supplies a continuous stream or spray of water at the point of impact.		
	– When used outdoors.	None	APF 10
	– When used indoors or in an enclosed area.	APF 10	APF 10
	OR		
	Use tool equipped with commercially available shroud and dust collection system.		
	Operate and maintain tool in accordance with manufacturer's instructions to minimize dust emissions.		
	Dust collector must provide the air flow recommended by the tool manufacturer, or greater, and have a filter with 99 percent or greater efficiency and a filter-cleaning mechanism.		
	– When used outdoors.	None	APF 10
	– When used indoors or in an enclosed area.	APF 10	APF 10
(xi) Handheld grinders for mortar removal (i.e., tuckpointing)	Use grinder equipped with commercially available shroud and dust collection system.	APF 10	APF 25
	Operate and maintain tool in accordance with manufacturer's instructions to minimize dust emissions.		
	Dust collector must provide 25 cubic feet per minute (cfm) or greater of airflow per inch of wheel diameter and have a filter with 99 percent or greater efficiency and a cyclonic preseparator or filter-cleaning mechanism.		

(Continued)

TABLE E.1 (CONTINUED)
Specified Exposure Control Methods When Working with Materials Containing Crystalline Silica

Equipment/Task	Engineering and Work Practice Control Methods	Required Respiratory Protection and Minimum Assigned Protection Factor (APF)	
		≤4 Hours/ Shift	>4 Hours/ Shift
(xii) Handheld grinders for uses other than mortar removal	For tasks performed outdoors only:		
	Use grinder equipped with integrated water delivery system that continuously feeds water to the grinding surface.	None	None
	Operate and maintain tool in accordance with manufacturer's instructions to minimize dust emissions.		
	OR		
	Use grinder equipped with commercially available shroud and dust collection system.		
	Operate and maintain tool in accordance with manufacturer's instructions to minimize dust emissions.		
	Dust collector must provide 25 cubic feet per minute (cfm) or greater of airflow per inch of wheel diameter and have a filter with 99 percent or greater efficiency and a cyclonic pre-separator or filter-cleaning mechanism.		
	– When used outdoors.	None	None
	– When used indoors or in an enclosed area.	None	APF10
(xiii) Walk-behind milling machines and floor grinders	Use machine equipped with integrated water delivery system that continuously feeds water to the cutting surface.	None	None
	Operate and maintain tool in accordance with manufacturer's instructions to minimize dust emissions.		
	OR		
	Use machine equipped with dust collection system recommended by the manufacturer.	None	None
	Operate and maintain tool in accordance with manufacturer's instructions to minimize dust emissions.		
	Dust collector must provide the air flow recommended by the manufacturer, or greater, and have a filter with 99 percent or greater efficiency and a filter-cleaning mechanism.		

(Continued)

TABLE E.1 (CONTINUED)
Specified Exposure Control Methods When Working with Materials Containing Crystalline Silica

Equipment/Task	Engineering and Work Practice Control Methods	Required Respiratory Protection and Minimum Assigned Protection Factor (APF)	
		≤4 Hours/ Shift	>4 Hours/ Shift
	When used indoors or in an enclosed area, use a HEPA-filtered vacuum to remove loose dust in between passes.		
(xiv) Small drivable milling machines (less than half-lane)	Use a machine equipped with supplemental water sprays designed to suppress dust. Water must be combined with a surfactant. Operate and maintain machine to minimize dust emissions.	None	None
(xv) Large drivable milling machines (half-lane and larger)	For cuts of any depth on asphalt only: Use machine equipped with exhaust ventilation on drum enclosure and supplemental water sprays designed to suppress dust. Operate and maintain machine to minimize dust emissions.	None	None
	For cuts of four inches in depth or less on any substrate: Use machine equipped with exhaust ventilation on drum enclosure and supplemental water sprays designed to suppress dust. Operate and maintain machine to minimize dust emissions.	None	None
	OR Use a machine equipped with supplemental water spray designed to suppress dust. Water must be combined with a surfactant. Operate and maintain machine to minimize dust emissions.	None	None
(xvi) Crushing machines	Use equipment designed to deliver water spray or mist for dust suppression at crusher and other points where dust is generated (e.g., hoppers, conveyers, sieves/sizing or vibrating components, and discharge points). Operate and maintain machine in accordance with manufacturer's instructions to minimize dust emissions.	None	None

(Continued)

TABLE E.1 (CONTINUED)
Specified Exposure Control Methods When Working with Materials
Containing Crystalline Silica

Equipment/Task	Engineering and Work Practice Control Methods	Required Respiratory Protection and Minimum Assigned Protection Factor (APF)	
		≤4 Hours/ Shift	>4 Hours/ Shift
	Use a ventilated booth that provides fresh, climate-controlled air to the operator, or a remote control station.		
(xvii) Heavy equipment and utility vehicles used to abrade or fracture silica-containing materials (e.g., hoe-ramming, rock ripping) or used during demolition activities involving silica-containing materials	Operate equipment from within an enclosed cab.	None	None
	When employees outside of the cab are engaged in the task, apply water and/or dust suppressants as necessary to minimize dust emissions.	None	None
(xviii) Heavy equipment and utility vehicles for tasks such as grading and excavating but not including: demolishing, abrading, or fracturing silica-containing materials	Apply water or dust suppressants as necessary to minimize dust emmissions. OR	None	None
	When the equipment operator is the only employee engaged in the task, operate equipment from within an enclosed cab.	None	None

1926.1153(c)(2)(iii)(C)
Has gaskets and seals that are in good condition and working properly.

1926.1153(c)(2)(iii)(D)
Is under positive pressure maintained through continuous delivery of fresh air.

1926.1153(c)(2)(iii)(E)
Has intake air that is filtered through a filter that is 95 percent efficient in the 0.3–10.0 μm range (e.g., MERV-16 or better).

1926.1153(c)(2)(iii)(F)
Has heating and cooling capabilities.

1926.1153(c)(3)

Where an employee performs more than one task on Table E.1 during the course of a shift, and the total duration of all tasks combined is more than four hours, the required respiratory protection for each task is the respiratory protection specified for more than four hours per shift. If the total duration of all tasks on Table E.1 combined is less than four hours, the required respiratory protection for each task is the respiratory protection specified for less than four hours per shift.

1926.1153(d)

Alternative exposure control methods. For tasks not listed in Table E.1, or where the employer does not fully and properly implement the engineering controls, work practices, and respiratory protection described in Table E.1:

1926.1153(d)(1)

Permissible exposure limit (PEL). The employer shall ensure that no employee is exposed to an airborne concentration of respirable crystalline silica in excess of 50 µg/m^3, calculated as an eight-hour TWA.

1926.1153(d)(2)

Exposure assessment.

1926.1153(d)(2)(i)

General. The employer shall assess the exposure of each employee who is or may reasonably be expected to be exposed to respirable crystalline silica at or above the action level in accordance with either the performance option in paragraph (d)(2)(ii) or the scheduled monitoring option in paragraph (d)(2)(iii) of this section.

1926.1153(d)(2)(ii)

Performance option. The employer shall assess the eight-hour TWA exposure for each employee on the basis of any combination of air monitoring data or objective data sufficient to accurately characterize employee exposures to respirable crystalline silica.

1926.1153(d)(2)(iii)

Scheduled monitoring option.

1926.1153(d)(2)(iii)(A)

The employer shall perform initial monitoring to assess the eight-hour TWA exposure for each employee on the basis of one or more personal breathing zone air samples that reflect the exposures of employees on each shift, for each job classification, in each work area. Where several employees perform the same tasks on the same shift and in the same work area, the employer may sample a representative fraction of these employees in order to meet this requirement. In representative sampling, the employer shall sample the employee(s) who are expected to have the highest exposure to respirable crystalline silica.

1926.1153(d)(2)(iii)(B)

If initial monitoring indicates that employee exposures are below the action level, the employer may discontinue monitoring for those employees whose exposures are represented by such monitoring.

1926.1153(d)(2)(iii)(C)
Where the most recent exposure monitoring indicates that employee exposures are at or above the action level but at or below the PEL, the employer shall repeat such monitoring within six months of the most recent monitoring.

1926.1153(d)(2)(iii)(D)
Where the most recent exposure monitoring indicates that employee exposures are above the PEL, the employer shall repeat such monitoring within three months of the most recent monitoring.

1926.1153(d)(2)(iii)(E)
Where the most recent (noninitial) exposure monitoring indicates that employee exposures are below the action level, the employer shall repeat such monitoring within six months of the most recent monitoring until two consecutive measurements, taken seven or more days apart, are below the action level, at which time the employer may discontinue monitoring for those employees whose exposures are represented by such monitoring, except as otherwise provided in paragraph (d)(2)(iv) of this section.

1926.1153(d)(2)(iv)
Reassessment of exposures. The employer shall reassess exposures whenever a change in the production, process, control equipment, personnel, or work practices may reasonably be expected to result in new or additional exposures at or above the action level, or when the employer has any reason to believe that new or additional exposures at or above the action level have occurred.

1926.1153(d)(2)(v)
Methods of sample analysis. The employer shall ensure that all samples taken to satisfy the monitoring requirements of paragraph (d)(2) of this section are evaluated by a laboratory that analyzes air samples for respirable crystalline silica in accordance with the procedures in Appendix A to this section.

1926.1153(d)(2)(vi)
Employee notification of assessment results.

1926.1153(d)(2)(vi)(A)
Within five working days after completing an exposure assessment in accordance with paragraph (d)(2) of this section, the employer shall individually notify each affected employee in writing of the results of that assessment or post the results in an appropriate location accessible to all affected employees.

1926.1153(d)(2)(vi)(B)
Whenever an exposure assessment indicates that employee exposure is above the PEL, the employer shall describe in the written notification the corrective action being taken to reduce employee exposure to or below the PEL.

1926.1153(d)(2)(vii)
Observation of monitoring.

1926.1153(d)(2)(vii)(A)
Where air monitoring is performed to comply with the requirements of this section, the employer shall provide affected employees or their designated representatives an

opportunity to observe any monitoring of employee exposure to respirable crystalline silica.

1926.1153(d)(2)(vii)(B)

When observation of monitoring requires entry into an area where the use of protective clothing or equipment is required for any workplace hazard, the employer shall provide the observer with protective clothing and equipment at no cost and shall ensure that the observer uses such clothing and equipment.

1926.1153(d)(3)

Methods of compliance.

1926.1153(d)(3)(i)

Engineering and work practice controls. The employer shall use engineering and work practice controls to reduce and maintain employee exposure to respirable crystalline silica to or below the PEL, unless the employer can demonstrate that such controls are not feasible. Wherever such feasible engineering and work practice controls are not sufficient to reduce employee exposure to or below the PEL, the employer shall nonetheless use them to reduce employee exposure to the lowest feasible level and shall supplement them with the use of respiratory protection that complies with the requirements of paragraph (e) of this section.

1926.1153(d)(3)(ii)

Abrasive blasting. In addition to the requirements of paragraph (d)(3)(i) of this section, the employer shall comply with other OSHA standards, when applicable, such as 29 CFR 1926.57 (Ventilation), where abrasive blasting is conducted using crystalline silica-containing blasting agents, or where abrasive blasting is conducted on substrates that contain crystalline silica.

1926.1153(e)

Respiratory protection.

1926.1153(e)(1)

General. Where respiratory protection is required by this section, the employer must provide each employee an appropriate respirator that complies with the requirements of this paragraph and 29 CFR 1910.134. Respiratory protection is required:

1926.1153(e)(1)(i)

Where specified by Table E.1 of paragraph (c) of this section or

1926.1153(e)(1)(ii)

For tasks not listed in Table E.1, or where the employer does not fully and properly implement the engineering controls, work practices, and respiratory protection described in Table E.1:

1926.1153(e)(1)(ii)(A)

Where exposures exceed the PEL during periods necessary to install or implement feasible engineering and work practice controls.

1926.1153(e)(1)(ii)(B)
Where exposures exceed the PEL during tasks, such as certain maintenance and repair tasks, for which engineering and work practice controls are not feasible.

1926.1153(e)(1)(ii)(C)
During tasks for which an employer has implemented all feasible engineering and work practice controls and such controls are not sufficient to reduce exposures to or below the PEL.

1926.1153(e)(2)
Respiratory protection program. Where respirator use is required by this section, the employer shall institute a respiratory protection program in accordance with 29 CFR 1910.134.

1926.1153(e)(3)
Specified exposure control methods. For the tasks listed in Table E.1 in paragraph (c) of this section, if the employer fully and properly implements the engineering controls, work practices, and respiratory protection described in Table E.1, the employer shall be considered to be in compliance with paragraph (e)(1) of this section and the requirements for selection of respirators in 29 CFR 1910.134(d)(1)(iii) and (d)(3) with regard to exposure to respirable crystalline silica.

1926.1153(f)
Housekeeping.

1926.1153(f)(1)
The employer shall not allow dry sweeping or dry brushing where such activity could contribute to employee exposure to respirable crystalline silica unless wet sweeping, HEPA-filtered vacuuming or other methods that minimize the likelihood of exposure are not feasible.

1926.1153(f)(2)
The employer shall not allow compressed air to be used to clean clothing or surfaces where such activity could contribute to employee exposure to respirable crystalline silica unless:

1926.1153(f)(2)(i)
The compressed air is used in conjunction with a ventilation system that effectively captures the dust cloud created by the compressed air or

1926.1153(f)(2)(ii)
No alternative method is feasible.

1926.1153(g)
Written exposure control plan.

1926.1153(g)(1)
The employer shall establish and implement a written exposure control plan that contains at least the following elements:

1926.1153(g)(1)(i)

A description of the tasks in the workplace that involve exposure to respirable crystalline silica.

1926.1153(g)(1)(ii)

A description of the engineering controls, work practices, and respiratory protection used to limit employee exposure to respirable crystalline silica for each task.

1926.1153(g)(1)(iii)

A description of the housekeeping measures used to limit employee exposure to respirable crystalline silica.

1926.1153(g)(1)(iv)

A description of the procedures used to restrict access to work areas, when necessary, to minimize the number of employees exposed to respirable crystalline silica and their level of exposure, including exposures generated by other employers or sole proprietors.

1926.1153(g)(2)

The employer shall review and evaluate the effectiveness of the written exposure control plan at least annually and update it as necessary.

1926.1153(g)(3)

The employer shall make the written exposure control plan readily available for examination and copying, upon request, to each employee covered by this section, their designated representatives, the assistant secretary and the director.

1926.1153(g)(4)

The employer shall designate a competent person to make frequent and regular inspections of job sites, materials, and equipment to implement the written exposure control plan.

1926.1153(h)

Medical surveillance.

1926.1153(h)(1)

General.

1926.1153(h)(1)(i)

The employer shall make medical surveillance available at no cost to the employee, and at a reasonable time and place, for each employee who will be required under this section to use a respirator for 30 or more days per year.

1926.1153(h)(1)(ii)

The employer shall ensure that all medical examinations and procedures required by this section are performed by a PLHCP as defined in paragraph (b) of this section.

1926.1153(h)(2)

Initial examination. The employer shall make available an initial (baseline) medical examination within 30 days after initial assignment, unless the employee has received a medical examination that meets the requirements of this section within the last three years. The examination shall consist of:

1926.1153(h)(2)(i)
A medical and work history, with emphasis on past, present, and anticipated exposure to respirable crystalline silica, dust, and other agents affecting the respiratory system; any history of respiratory system dysfunction, including signs and symptoms of respiratory disease (e.g., shortness of breath, coughing, wheezing); history of tuberculosis; and smoking status and history.

1926.1153(h)(2)(ii)
A physical examination with special emphasis on the respiratory system.

1926.1153(h)(2)(iii)
A chest x-ray (a single posteroanterior radiographic projection or radiograph of the chest at full inspiration recorded on either film (no less than 14 × 17 inches and no more than 16 × 17 inches) or digital radiography systems), interpreted and classified according to the International Labor Office (ILO) International Classification of Radiographs of Pneumoconiosis by a NIOSH-certified B Reader.

1926.1153(h)(2)(iv)
A pulmonary function test to include forced vital capacity (FVC) and forced expiratory volume in one second (FEV1) and FEV1/FVC ratio, administered by a spirometry technician with a current certificate from a NIOSHapproved spirometry course.

1926.1153(h)(2)(v)
Testing for latent tuberculosis infection.

1926.1153(h)(2)(vi)
Any other tests deemed appropriate by the PLHCP.

1926.1153(h)(3)
Periodic examinations. The employer shall make available medical examinations that include the procedures described in paragraph (h)(2) of this section (except paragraph (h)(2)(v)) at least every three years, or more frequently if recommended by the PLHCP.

1926.1153(h)(4)
Information provided to the PLHCP. The employer shall ensure that the examining PLHCP has a copy of this standard, and shall provide the PLHCP with the following information:

1926.1153(h)(4)(i)
A description of the employee's former, current, and anticipated duties as they relate to the employee's occupational exposure to respirable crystalline silica.

1926.1153(h)(4)(ii)
The employee's former, current, and anticipated levels of occupational exposure to respirable crystalline silica.

1926.1153(h)(4)(iii)
A description of any personal protective equipment used or to be used by the employee, including when and for how long the employee has used or will use that equipment.

1926.1153(h)(4)(iv)
Information from records of employment-related medical examinations previously provided to the employee and currently within the control of the employer.

1926.1153(h)(5)
PLHCP's written medical report for the employee. The employer shall ensure that the PLHCP explains to the employee the results of the medical examination and provides each employee with a written medical report within 30 days of each medical examination performed. The written report shall contain:

1926.1153(h)(5)(i)
A statement indicating the results of the medical examination, including any medical condition(s) that would place the employee at increased risk of material impairment to health from exposure to respirable crystalline silica and any medical conditions that require further evaluation or treatment.

1926.1153(h)(5)(ii)
Any recommended limitations on the employee's use of respirators.

Appendix F: Fall Protection Inspection Guidelines and New OSHA General Industry Walking–Working Surfaces and Fall Protection Standards

FALL PROTECTION INSPECTION GUIDELINES

INSPECTION AND MAINTENANCE

To maintain their service life and high performance, all belts and harnesses should be inspected frequently. Visual inspection before each use should become routine, as well as routine inspection by a competent person. If any of the conditions listed below are found, the equipment should be replaced before being used.

HARNESS INSPECTION

BELTS AND RINGS

For harness inspections, begin at one end and hold the body side of the belt toward you, grasping the belt with your hands six to eight inches apart. Bend the belt in an inverted U. Watch for frayed edges, broken fibers, pulled stitches, cuts, or chemical damage. Check D-rings and D-ring metal wear pads for distortion, cracks, breaks, and rough or sharp edges. The D-ring bar should be at a 90° angle to the long axis of the belt and should pivot freely. Attachments of buckles and D-rings should be given special attention. Note any unusual wear, frayed or cut fibers, or distortion of the buckles. Rivets should be tight and not be removable with fingers. The body-side rivet base and outside rivets should be flat against the material. Bent rivets will fail under stress. Inspect frayed or broken strands. Broken webbing strands generally appear as tufts on the webbing surface. Any broken, cut, or burned stitches will be readily seen.

Tongue Buckle

Buckle tongues should be free of distortion in shape and motion. They should overlap the buckle frame and move freely back and forth in their socket. Rollers should turn freely on the frame. Check for distortion or sharp edges.

Friction Buckle

Inspect the buckle for distortion. The outer bar or center bars must be straight. Pay special attention to corners and attachment points of the center bar.

LANYARD INSPECTION

When inspecting lanyards, begin at one end and work to the opposite end. Slowly rotate the lanyard so that the entire circumference is checked. Spliced ends require particular attention. Hardware should be examined using the procedures detailed below.

Hardware

Snaps

Inspect closely for hook and eye distortion, cracks, corrosion, or pitted surfaces. The keeper or latch should seat into the nose without binding and should not be distorted or obstructed. The keeper spring should exert sufficient force to close the keeper firmly. Keeper rocks must prevent the keeper from opening when the keeper is closed.

Thimbles

The thimble (protective plastic sleeve) must be firmly seated in the eye of the splice, and the splice should have no loose or cut strands. The edges of the thimble should be free of sharp edges, distortion, or cracks.

Lanyards

Steel Lanyards

While rotating a steel lanyard, watch for cuts, frayed areas, or unusual wear patterns on the wire. The use of steel lanyards for fall protection without a shock-absorbing device is not recommended.

Web Lanyards

While bending webbing over a piece of pipe, observe each side of the webbed lanyard. This will reveal any cuts or breaks. Due to the limited elasticity of the web lanyard, fall protection without the use of a shock absorber is not recommended.

Rope Lanyards

Rotation of the rope lanyard while inspecting it from end to end will bring to light any fuzzy, worn, broken, or cut fibers. Weakened areas from extreme loads will appear as a noticeable change in original diameter. The rope diameter should be uniform throughout, following a short break-in period. When a rope lanyard is used for fall protection, a shock-absorbing system should be included.

VISUAL INDICATION OF DAMAGE TO WEBBING AND ROPE LANYARDS

Heat

In excessive heat, nylon becomes brittle and has a shriveled, brownish appearance. Fibers will break when flexed and should not be used above 180°F.

Chemical

Change in color usually appears as a brownish smear or smudge. Transverse cracks, which appear when belt is bent over tight, causes a loss of elasticity in the belt.

Ultraviolet Rays

Do not store webbing and rope lanyards in direct sunlight, because ultraviolet rays can reduce the strength of some material.

Molten Metal or Flame

Webbing and rope strands may be fused together by molten metal or flame. Watch for hard, shiny spots or a hard and brittle feel. Webbing will not support combustion, but nylon will.

Paint and Solvents

Paint will penetrate and dry, restricting movements of fibers. Drying agents and solvents in some paints will appear as chemical damage.

SHOCK-ABSORBING PACKS

The outer portion of the shock-absorbing pack should be examined for burn holes and tears. Stitching on areas where the pack is sewn to the D-ring, belt, or lanyard should be examined for loose strands, rips, and deterioration.

CLEANING OF EQUIPMENT

Basic care for fall-protection safety equipment will prolong and endure the life of the equipment and contribute toward the performance of its vital safety function. Proper storage and maintenance after use are as important as cleaning the equipment of dirt, corrosives, or contaminants. The storage area should be clean, dry, and free of exposure to fumes or corrosive elements.

Nylon and Polyester

Wipe off all surface dirt with a sponge dampened in plain water. Squeeze the sponge dry. Dip the sponge in a mild solution of water and commercial soap or detergent. Work up a thick lather with a vigorous back-and-forth motion, then wipe the belt dry with a clean cloth. Hang freely to dry but away from excessive heat.

Drying

Harnesses, belts, and other equipment should be dried thoroughly without exposure to heat, steam, or long periods of sunlight.

29 CFR 1910.28

1910.28(a)
General.

1910.28(a)(1)
This section requires employers to provide protection for each employee exposed to fall and falling object hazards. Unless stated otherwise, the employer must ensure that all fall protection and falling object protection required by this section meet the criteria in § 1910.29, except that personal fall-protection systems required by this section meet the criteria of § 1910.140.

1910.28(a)(2)
This section does not apply:

1910.28(a)(2)(i)
To portable ladders.

1910.28(a)(2)(ii)
When employers are inspecting, investigating, or assessing workplace conditions or work to be performed prior to the start of work or after all work has been completed. This exemption does not apply when fall-protection systems or equipment meeting the requirements of § 1910.29 have been installed and are available for workers to use for prework and postwork inspections, investigations, or assessments.

1910.28(a)(2)(iii)
To fall hazards presented by the exposed perimeters of entertainment stages and the exposed perimeters of rail-station platforms.

1910.28(a)(2)(iv)
To powered platforms covered by § 1910.66(j).

1910.28(a)(2)(v)
To aerial lifts covered by § 1910.67(c)(2)(v).

1910.28(a)(2)(vi)
To telecommunications work covered by § 1910.268(n)(7) and (8).

1910.28(a)(2)(vii)

To electric power generation, transmission, and distribution work covered by § 1910.269(g)(2)(i).

1910.28(b)
Protection from fall hazards.

1910.28(b)(1)
Unprotected sides and edges.

1910.28(b)(1)(i)

Except as provided elsewhere in this section, the employer must ensure that each employee on a walking–working surface with an unprotected side or edge that is four feet (1.2 meters) or more above a lower level is protected from falling by one or more of the following:

1910.28(b)(1)(i)(A)
Guardrail systems.

1910.28(b)(1)(i)(B)
Safety net systems.

1910.28(b)(1)(i)(C)
Personal fall-protection systems, such as personal fall-arrest, travel restraint, or positioning systems.

1910.28(b)(1)(ii)

When the employer can demonstrate that it is not feasible or creates a greater hazard to use guardrail, safety net, or personal fall-protection systems on residential roofs, the employer must develop and implement a fall protection plan that meets the requirements of 29 CFR 1926.502(k) and training that meets the requirements of 29 CFR 1926.503(a) and (c).

Note to paragraph (b)(1)(ii) of this section: There is a presumption that it is feasible and will not create a greater hazard to use at least one of the above-listed fall-protection systems specified in paragraph (b)(1)(i) of this section. Accordingly, the employer has the burden of establishing that it is not feasible or creates a greater hazard to provide the fall-protection systems specified in paragraph (b)(1)(i) and that it is necessary to implement a fall protection plan that complies with § 1926.502(k) in the particular work operation, in lieu of implementing any of those systems.

1910.28(b)(1)(iii)

When the employer can demonstrate that the use of fall-protection systems is not feasible on the working side of a platform used at a loading rack, loading dock, or teeming platform, the work may be done without a fall-protection system, provided:

1910.28(b)(1)(iii)(A)
The work operation for which fall protection is infeasible is in process.

1910.28(b)(1)(iii)(B)
Access to the platform is limited to authorized employees.

1910.28(b)(1)(iii)(C)
The authorized employees are trained in accordance with § 1910.30.

1910.28(b)(2)
Hoist areas. The employer must ensure:

1910.28(b)(2)(i)
Each employee in a hoist area is protected from falling four feet (1.2 meters) or more to a lower level by

1910.28(b)(2)(i)(A)
A guardrail system.

1910.28(b)(2)(i)(B)
A personal fall-arrest system.

1910.28(b)(2)(i)(C)
A travel restraint system.

1910.28(b)(2)(ii)
When any portion of a guardrail system, gate, or chains is removed, and an employee must lean through or over the edge of the access opening to facilitate hoisting, the employee is protected from falling by a personal fall-arrest system.

1910.28(b)(2)(iii)
If grab handles are installed at hoist areas, they meet the requirements of § 1910.29(l).

1910.28(b)(3)
Holes. The employer must ensure:

1910.28(b)(3)(i)
Each employee is protected from falling through any hole (including skylights) that is four feet (1.2 meters) or more above a lower level by one or more of the following:

1910.28(b)(3)(i)(A)
Covers.

1910.28(b)(3)(i)(B)
Guardrail systems.

1910.28(b)(3)(i)(C)
Travel restraint systems.

1910.28(b)(3)(i)(D)
Personal fall-arrest systems.

1910.28(b)(3)(ii)
Each employee is protected from tripping into or stepping into or through any hole that is less than four feet (1.2 meters) above a lower level by covers or guardrail systems.

1910.28(b)(3)(iii)
Each employee is protected from falling into a stairway floor hole by a fixed guardrail system on all exposed sides, except at the stairway entrance. However, for any

stairway used less than once per day where traffic across the stairway floor hole prevents the use of a fixed guardrail system (e.g., holes located in aisle spaces), the employer may protect employees from falling into the hole by using a hinged floor hole cover that meets the criteria in § 1910.29 and a removable guardrail system on all exposed sides, except at the entrance to the stairway.

1910.28(b)(3)(iv)
Each employee is protected from falling into a ladderway floor hole or ladderway platform hole by a guardrail system and toeboards erected on all exposed sides, except at the entrance to the hole, where a self-closing gate or an offset must be used.

1910.28(b)(3)(v)
Each employee is protected from falling through a hatchway and chutefloor hole by

1910.28(b)(3)(v)(A)
A hinged floor-hole cover that meets the criteria in § 1910.29 and a fixed guardrail system that leaves only one exposed side. When the hole is not in use, the employer must ensure the cover is closed or a removable guardrail system is provided on the exposed sides.

1910.28(b)(3)(v)(B)
A removable guardrail system and toeboards on not more than two sides of the hole and a fixed guardrail system on all other exposed sides. The employer must ensure the removable guardrail system is kept in place when the hole is not in use.

1910.28(b)(3)(v)(C)
A guardrail system or a travel restraint system when a work operation necessitates passing material through a hatchway or chute floor hole.

1910.28(b)(4)
Dockboards.

1910.28(b)(4)(i)
The employer must ensure that each employee on a dockboard is protected from falling four feet (1.2 meters) or more to a lower level by a guardrail system or handrails.

1910.28(b)(4)(ii)
A guardrail system or handrails are not required when

1910.28(b)(4)(ii)(A)
Dockboards are being used solely for materials-handling operations using motorized equipment.

1910.28(b)(4)(ii)(B)
Employees engaged in these operations are not exposed to fall hazards greater than 10 feet (3 meters).

1910.28(b)(4)(ii)(C)
Those employees have been trained in accordance with § 1910.30.

1910.28(b)(5)
Runways and similar walkways.

1910.28(b)(5)(i)

The employer must ensure each employee on a runway or similar walkway is protected from falling four feet (1.2 meters) or more to a lower level by a guardrail system.

1910.28(b)(5)(ii)

When the employer can demonstrate that it is not feasible to have guardrails on both sides of a runway used exclusively for a special purpose, the employer may omit the guardrail on one side of the runway, provided the employer ensures:

1910.28(b)(5)(ii)(A)

The runway is at least 18 inches (46 centimeters) wide.

1910.28(b)(5)(ii)(B)

Each employee is provided with and uses a personal fall-arrest system or travel restraint system.

1910.28(b)(6)

Dangerous equipment. The employer must ensure:

1910.28(b)(6)(i)

Each employee less than four feet (1.2 meters) above dangerous equipment is protected from falling into or onto the dangerous equipment by a guardrail system or a travel restraint system, unless the equipment is covered or guarded to eliminate the hazard.

1910.28(b)(6)(ii)

Each employee four feet (1.2 meters) or more above dangerous equipment must be protected from falling by

1910.28(b)(6)(ii)(A)

Guardrail systems.

1910.28(b)(6)(ii)(B)

Safety net systems.

1910.28(b)(6)(ii)(C)

Travel restraint systems.

1910.28(b)(6)(ii)(D)

Personal fall-arrest systems.

1910.28(b)(7)

Openings. The employer must ensure that each employee on a walking–working surface near an opening, including one with a chute attached, where the inside bottom edge of the opening is less than 39 inches (99 centimeters) above that walking–working surface and the outside bottom edge of the opening is four feet (1.2 meters) or more above a lower level is protected from falling by the use of

1910.28(b)(7)(i)

Guardrail systems.

1910.28(b)(7)(ii)
Safety net systems.

1910.28(b)(7)(iii)
Travel restraint systems.

1910.28(b)(7)(iv)
Personal fall-arrest systems.

1910.28(b)(8)
Repair pits, service pits, and assembly pits less than 10 feet in depth. The use of a fall-protection system is not required for a repair pit, service pit, or assembly pit that is less than 10 feet (3 meters) deep, provided the employer:

1910.28(b)(8)(i)
Limits access within six feet (1.8 meters) of the edge of the pit to authorized employees trained in accordance with § 1910.30.

1910.28(b)(8)(ii)
Applies floor markings at least six feet (1.8 meters) from the edge of the pit in colors that contrast with the surrounding area; or places a warning line at least six feet (1.8 meters) from the edge of the pit as well as stanchions that are capable of resisting, without tipping over, a force of at least 16 pounds (71 N) applied horizontally against the stanchion at a height of 30 inches (76 centimeters); or places a combination of floor markings and warning lines at least six feet (1.8 meters) from the edge of the pit. When two or more pits in a common area are not more than 15 feet (4.5 meters) apart, the employer may comply by placing contrasting floor markings at least six feet (1.8 meters) from the pit edge around the entire area of the pits.

1910.28(b)(8)(iii)
Posts readily visible caution signs that meet the requirements of § 1910.145 and state "Caution: Open Pit."

1910.28(b)(9)
Fixed ladders (that extend more than 24 feet [7.3 meters] above a lower level).

1910.28(b)(9)(i)
For fixed ladders that extend more than 24 feet (7.3 meters) above a lower level, the employer must ensure:

1910.28(b)(9)(i)(A)
Existing fixed ladders. Each fixed ladder installed before November 19, 2018, is equipped with a personal fall-arrest system, ladder safety system, cage, or well.

1910.28(b)(9)(i)(B)
New fixed ladders. Each fixed ladder installed on and after November 19, 2018, is equipped with a personal fall-arrest system or a ladder safety system.

1910.28(b)(9)(i)(C)
Replacement. When a fixed ladder, cage, or well, or any portion of a section thereof, is replaced, a personal fall-arrest system or ladder safety system is installed in at least that section of the fixed ladder, cage, or well where the replacement is located.

1910.28(b)(9)(i)(D)
Final deadline. On and after November 18, 2036, all fixed ladders are equipped with a personal fall-arrest system or a ladder safety system.

1910.28(b)(9)(ii)
When a one-section fixed ladder is equipped with a personal fall protection or a ladder safety system or a fixed ladder is equipped with a personal fall-arrest or ladder safety system on more than one section, the employer must ensure:

1910.28(b)(9)(ii)(A)
The personal fall-arrest system or ladder safety system provides protection throughout the entire vertical distance of the ladder, including all ladder sections.

1910.28(b)(9)(ii)(B)
The ladder has rest platforms provided at maximum intervals of 150 feet (45.7 meters).

1910.28(b)(9)(iii)
The employer must ensure ladder sections having a cage or well:

1910.28(b)(9)(iii)(A)
Are offset from adjacent sections.

1910.28(b)(9)(iii)(B)
Have landing platforms provided at maximum intervals of 50 feet (15.2 meters).

1910.28(b)(9)(iv)
The employer may use a cage or well in combination with a personal fall-arrest system or ladder safety system provided that the cage or well does not interfere with the operation of the system.

1910.28(b)(10)
Outdoor advertising (billboards).

1910.28(b)(10)(i)
The requirements in paragraph (b)(9) of this section, and other requirements in subparts D and I of this part, apply to fixed ladders used in outdoor advertising activities.

1910.28(b)(10)(ii)
When an employee engaged in outdoor advertising climbs a fixed ladder before November 19, 2018, that is not equipped with a cage, well, personal fall-arrest system, or a ladder safety system the employer must ensure the employee:

1910.28(b)(10)(ii)(A)
Receives training and demonstrates the physical capability to perform the necessary climbs in accordance with § 1910.29(h).

1910.28(b)(10)(ii)(B)
Wears a body harness equipped with an 18-inch (46-centimeter) rest lanyard.

1910.28(b)(10)(ii)(C)
Keeps both hands free of tools or material when climbing on the ladder.

1910.28(b)(10)(ii)(D)
Is protected by a fall-protection system upon reaching the work position.

1910.28(b)(11)
Stairways. The employer must ensure:

1910.28(b)(11)(i)
Each employee exposed to an unprotected side or edge of a stairway landing that is four feet (1.2 meters) or more above a lower level is protected by a guardrail or stair rail system.

1910.28(b)(11)(ii)
Each flight of stairs having at least three treads and at least four risers is equipped with stair rail systems and handrails as follows.

1910.28(b)(11)(iii)
Each ship stairs and alternating tread type stairs is equipped with handrails on both sides.

1910.28(b)(12)
Scaffolds and rope descent systems. The employer must ensure:

1910.28(b)(12)(i)
Each employee on a scaffold is protected from falling in accordance 29 CFR part 1926, subpart L.

1910.28(b)(12)(ii)
Each employee using a rope descent system four feet (1.2 meters) or more above a lower level is protected from falling by a personal fall-arrest system.

1910.28(b)(13)
Work on low-slope roofs.

1910.28(b)(13)(i)
When work is performed less than six feet (1.6 meters) from the roof edge, the employer must ensure each employee is protected from falling by a guardrail system, safety net system, travel restraint system, or personal fall-arrest system.

1910.28(b)(13)(ii)
When work is performed at least six feet (1.6 meters) but less than 15 feet (4.6 meters) from the roof edge, the employer must ensure each employee is protected from falling by using a guardrail system, safety net system, travel restraint system, or personal fall-arrest system. The employer may use a designated area when performing work that is both infrequent and temporary.

1910.28(b)(13)(iii)
When work is performed 15 feet (4.6 meters) or more from the roof edge, the employer must

1910.28(b)(13)(iii)(A)
Protect each employee from falling by a guardrail system, safety net system, travel restraint system, or personal fall-arrest system or a designated area. The employer is

not required to provide any fall protection, provided the work is both infrequent and temporary.

1910.28(b)(13)(iii)(B)
Implement and enforce a work rule prohibiting employees from going within 15 feet (4.6 meters) of the roof edge without using fall protection in accordance with paragraphs (b)(13)(i) and (ii) of this section.

1910.28(b)(14)
Slaughtering facility platforms.

1910.28(b)(14)(i)
The employer must protect each employee on the unprotected working side of a slaughtering facility platform that is four feet (1.2 meters) or more above a lower level from falling by using:

1910.28(b)(14)(i)(A)
Guardrail systems or

1910.28(b)(14)(i)(B)
Travel restraint systems.

1910.28(b)(14)(ii)
When the employer can demonstrate the use of a guardrail or travel restraint system is not feasible, the work may be done without those systems provided:

1910.28(b)(14)(ii)(A)
The work operation for which fall protection is infeasible is in process.

1910.28(b)(14)(ii)(B)
Access to the platform is limited to authorized employees.

1910.28(b)(14)(ii)(C)
The authorized employees are trained in accordance with § 1910.30.

1910.28(b)(15)
Walking–working surfaces not otherwise addressed. Except as provided elsewhere in this section or by other subparts of this part, the employer must ensure each employee on a walking–working surface four feet (1.2 meters) or more above a lower level is protected from falling by

1910.28(b)(15)(i)
Guardrail systems.

1910.28(b)(15)(ii)
Safety net systems.

1910.28(b)(15)(iii)
Personal fall-protection systems, such as personal fall-arrest, travel restraint, or positioning systems.

TABLE F.1
Stairway Handrail Requirements

Stair Width	Enclosed	One Open Side	Two Open Sides	With Earth Built Up on Both Sides
Less than 44 inches (1.1 m).	At least one handrail.	One stair rail system with handrail on open side.	One stair rail system each open side.	
44 inches (1.1 m) to 88 inches (2.2 m).	One handrail on each enclosed side.	One Stair rail system with handrail on open side and one handrail on enclosed side.	One stair rail system with handrail on each open side.	
Greater than 88 inches (2.2 m).	One handrail on each enclosed side and one intermediate handrail located in the middle of the stairs.	One stair rail system with handrail on open side, one handrail on enclosed side, and one intermediate handrail located in the middle of the stairs.	One stair rail system with handrail on each open side and one intermediate handrail located in the middle of the stairs.	
Exterior stairs less than 4.4 inches (1.1 m).				One handrail on least one side.

Note: The width of the stairs must be clear of all obstructions except handrails.

1910.28(c)

Protection from falling objects. When an employee is exposed to falling objects, the employer must ensure that each employee wears head protection that meets the requirements of subpart I of this part. In addition, the employer must protect employees from falling objects by implementing one or more of the following:

1910.28(c)(1)

Erecting toeboards, screens, or guardrail systems to prevent objects from falling to a lower level.

1910.28(c)(2)

Erecting canopy structures and keeping potential falling objects far enough from an edge, hole, or opening to prevent them from falling to a lower level.

1910.28(c)(3)

Barricading the area into which objects could fall, prohibiting employees from entering the barricaded area, and keeping objects far enough from an edge or opening to prevent them from falling to a lower level.

[39 FR 23502, June 27, 1974, as amended at 43 FR 49746, Oct. 24, 1978; 49 FR 5321, Feb. 10, 1984; 53 FR 12121, Apr. 12, 1988; 81 FR 82991-82994, Nov. 18, 2016]

29 CFR § 1910.29

Appendix G: OSHA Regulation, Citations, Compliance, and Rights/Responsibilities under the Act

Safety professionals should be aware of their rights and responsibilities when addressing the Occupational Safety and Health Administration (OSHA) or other governmental agencies before and after a disaster situation. Additionally, it is important that the safety professional be aware of the underlying purpose of the investigation and the potential civil and criminal penalties that can be assessed against the organization, as well as the safety professional or other members of the management team individually.

To begin with, the OSH Act covers virtually every American workplace that employs one or more employees and engages in a business that in any way affects interstate commerce.[1] The OSH Act covers employment in every state, the District of Columbia, Puerto Rico, Guam, the Virgin Islands, American Samoa, and the Trust Territory of the Pacific Islands.[2] The OSH Act does not, however, cover employees in situations where other state or federal agencies have jurisdiction that requires the agencies to prescribe or enforce their own safety and health regulations.[3] Additionally, the OSH Act exempts residential owners who employ people for ordinary domestic tasks, such as cooking, cleaning, and child care.[4] It also does not cover federal,[5] state, and local governments[6] or Native American reservations.[7]

The OSH Act does require every employer engaged in interstate commerce to furnish employees "a place of employment ... free from recognized hazards that are causing, or are likely to cause, death or serious harm."[8] To help employers create and maintain safe working environments and to enforce laws and regulations that ensure safe and healthful work environments, Congress provided for the creation of the Occupational Safety and Health Administration to be a new agency under the direction of the Department of Labor. Today, OSHA is one of the most widely known and powerful enforcement agencies. It has been granted broad regulatory powers to promulgate regulations and standards, investigate and inspect, issue citations, and propose penalties for safety violations in the workplace.

The OSH Act also established an independent agency to review OSHA citations and decisions, the Occupational Safety and Health Review Commission (OSHRC). The OSHRC is a quasi-judicial and independent administrative agency composed of three commissioners appointed by the president who serve staggered six-year terms. The OSHRC has the power to issue orders; uphold, vacate, or modify OSHA citations and penalties; and direct other appropriate relief and penalties.

The educational arm of the OSH Act is the National Institute for Occupational Safety and Health (NIOSH), which was created as a specialized educational agency of the existing National Institutes of Health. NIOSH conducts occupational safety and health research and develops criteria for new OSHA standards. NIOSH can conduct workplace inspections, issue subpoenas, and question employees and employers but it does not have the power to issue citations or penalties.

As permitted under the OSH Act, OSHA encourages individual states to take responsibility for OSHA administration and enforcement within their own respective boundaries. Each state possesses the ability to request and be granted the right to adopt state safety and health regulations and enforcement mechanisms.[9] For a state plan to be placed into effect, the state must first develop and submit its proposed program to the Secretary of Labor for review and approval. The secretary must certify that the state plan's standards are "at least as effective" as the federal standards and that the state will devote adequate resources to administering and enforcing standards.[10]

In most state plans, the state agency has developed more stringent safety and health standards than OSHA[11] and has usually developed more stringent enforcement schemes.[12] The Secretary of Labor has no statutory authority to reject a state plan if the proposed standards or enforcement scheme are more strict than the OSHA standards, but can reject the state plan if the standards are below the minimum limits set under OSHA standards.[13] These states are known as "state plan" states and territories.[14] Currently, there are twenty-six states plus Puerto Rico, and the Virgin Islands that have OSHA-approved state plans. Twenty-two state plans (and one U.S. territory) cover both private industry as well as state and local government workplaces. The remaining six five state plans and one U.S. territory plan cover only state and local government workers, and Federal OSHA has oversight of private industry. Employers in state plan states and territories must comply with their particular state's regulations; federal OSHA plays virtually no role in direct enforcement.

The Occupational Safety and Health Administration possess an approval and oversight role in regard to state plan programs. OSHA must approve all state plan proposals prior to enactment, and they maintain oversight authority to "pull the ticket" of any or all state plan programs at any time they are not achieving the identified prerequisites. Enforcement of this oversight authority was recently observed following a fire resulting in several workplace fatalities at the Imperial Foods facility in Hamlet, North Carolina. Following this incident, federal OSHA assumed jurisdiction and control over the state plan program in North Carolina and made significant modifications to this program before returning the program to state control.

The OSH Act requires that a covered employer comply with specific occupational safety and health standards and all rules, regulations, and orders issued pursuant to the OSH Act that apply to the workplace.[16] The OSH Act also requires that all standards be based on research, demonstration, experimentation, or other appropriate information.[17] The Secretary of Labor is authorized under the act to "promulgate, modify, or revoke any occupational safety and health standard,"[18] and the OSH Act describes the procedures that the secretary must follow when establishing new occupational safety and health standards.[19]

The OSH Act authorizes three ways to promulgate new standards. From 1970 to 1973, the Secretary of Labor was authorized in Section 6(a) of the Act[20] to adopt national consensus standards and establish federal safety and health standards without following lengthy rulemaking procedures. Many of the early OSHA standards were adapted mainly from other areas of regulation, such as the National Electric Code and American National Standards Institute (ANSI) guidelines; however, this promulgation method is no longer in effect.

The usual method of issuing, modifying, or revoking a new or existing OSHA standard is set out in Section 6(b) of the OSH Act and is known as informal rulemaking. It requires notice to interested parties, through subscription to the *Federal Register*, of the proposed regulation and standard and provides an opportunity for comment in a nonadversarial administrative hearing.[21] The proposed standard can also be advertised through magazine articles and other publications, thus informing interested parties of the proposed standard and regulation. This method differs from the requirements of most other administrative agencies that follow the Administrative Procedure Act[22] in that the OSH Act provides interested persons an opportunity to request a public hearing with oral testimony. It also requires the Secretary of Labor to publish in the *Federal Register* a notice of the time and place of such hearings.

Although not required under the OSH Act, the Secretary of Labor has directed, by regulation, that OSHA follow a more rigorous procedure for comment and hearing than other administrative agencies.[23] Upon notice and request for a hearing, OSHA must provide a hearing examiner in order to listen to any oral testimony offered. All oral testimony is preserved in a verbatim transcript. Interested persons are provided an opportunity to cross-examine OSHA representatives or others on critical issues. The secretary must state the reasons for the action to be taken on the proposed standard, and the statement must be supported by substantial evidence in the record as a whole.

The Secretary of Labor has the authority not to permit oral hearings and to call for written comment only. Within sixty days after the period for written comment or oral hearings has expired, the secretary must decide whether to adopt, modify, or revoke the standard in question. The secretary can also decide not to adopt a new standard. The secretary must then publish a statement of the reasons for any decision in the *Federal Register*. OSHA regulations further mandate that the secretary provide a supplemental statement of significant issues in the decision. Safety and health professionals should be aware that the standard as adopted and published in the *Federal Register* may be different from the proposed standard. The secretary is not required to reopen hearings when the adopted standard is a "logical outgrowth" of the proposed standard.[24]

The final method for promulgating new standards, and the one most infrequently used, is the emergency temporary standard permitted under Section 6(c).[25] The Secretary of Labor may establish a standard immediately if it is determined that employees are subjected to grave danger from exposure to substances or agents known to be toxic or physically harmful and that an emergency standard would protect the employees from the danger. An emergency temporary standard becomes effective on publication in the *Federal Register* and may remain in effect for six months. During this six-month period, the secretary must adopt a new permanent standard or abandon the emergency standard.

Only the Secretary of Labor can establish new OSHA standards. Recommendations or requests for an OSHA standard can come from any interested person or organization, including employees, employers, labor unions, environmental groups, and others.[26] When the secretary receives a petition to adopt a new standard or to modify or revoke an existing standard, he or she usually forwards the request to NIOSH and the National Advisory Committee on Occupational Safety and Health (NACOSH)[27] or the secretary may use a private organization such as the American National Standards Institute (ANSI) for advice and review.

The OSH Act requires that an employer must maintain a place of employment free from recognized hazards that are causing or are likely to cause death or serious physical harm, even if there is no specific OSHA standard addressing the circumstances. Under Section 5(a)(1), known as the "general duty clause (GDC)," an employer may be cited for a violation of the OSH Act if the condition causes harm or is likely to cause harm to employees, even if OSHA has not promulgated a standard specifically addressing the particular hazard. The general duty clause is a catchall standard encompassing all potential hazards that have not been specifically addressed in the OSHA standards. For example, if a company is cited for an ergonomic hazard and there is no ergonomic standard to apply, the hazard will be cited under the general duty clause.

Court precedent has established essential elements that must be present in order for OSHA to issue a GDC citation and prevail. These essential elements include the following:

1. The employer failed to keep the workplace free of a hazard to which employees of that employer were exposed.
2. The hazard was recognized. (Recognition may be established by employee knowledge, safety or management personnel knowledge, or trade/industry customs or standards.)
3. The hazard was causing or was likely to cause death of serious physical harm.
4. There was a feasible and effective/useful method to correct the hazard.
5. A specific OSHA rule/regulation does not exist for the hazard.

The OSH Act provides for a wide range of penalties, from a simple notice with no fine to criminal prosecution. OSHA penalties remained static from 1990 until 2016, and then penalties increased substantially as result of the Federal Civil Penalties Inflation Adjustment Act Improvements Act of 2015. This act mandated OSHA to increase its penalties in 2016 and then annually thereafter to account for inflation. On January 13, 2017, OSHA increased penalties accordingly, and Table G.1 represents OSHA penalties as they exist at the time of this book's publication.

Each alleged violation is categorized and the appropriate fine issued by the OSHA area director. It should be noted that each citation is separate and may carry with it a monetary fine. The gravity of the violation is the primary factor in determining penalties.[28] In assessing the gravity of a violation, the compliance officer or area

TABLE G.1

Type of Violation	Maximum Penalty
Serious	$12,675
Other-than-serious	
Posting	
Failure to Abate	$12,675 per day beyond date of abatement
Willful	$126,749
Repeat	

director must consider (1) the severity of the injury or illness that could result and (2) the probability that an injury or illness could occur as a result of the violation.[29] Specific penalty assessment tables assist the area director or compliance officer in determining the appropriate fine for the violation.[30]

After selecting the appropriate penalty table, the area director or other official determines the degree of probability that the injury or illness will occur by considering:[31]

1. The number of employees exposed.
2. The frequency and duration of the exposure.
3. The proximity of employees to the point of danger.
4. Factors such as the speed of the operation that require work under stress.
5. Other factors that might significantly affect the degree of probability of an accident.

In assessing monetary penalties, the area or regional director must consider the good faith of the employer, the gravity of the violation, the employer's past history of compliance, and the size of the employer.

OSHA has defined a serious violation as "an infraction in which there is a substantial probability that death or serious harm could result … unless the employer did not or could not with the exercise of reasonable diligence, know of the presence of the violation." Currently, the greatest monetary liabilities are for "repeat violations," "willful violations," and "failure to abate" cited violations. A repeat violation is a second citation for a violation that was cited previously by a compliance officer and the citation was part of a final order by the OSHRC. OSHA maintains records of all violations and must check for repeat violations after each inspection.

A willful violation is the employer's purposeful or negligent failure to correct a known deficiency. This type of violation, in addition to carrying a large monetary fine, exposes the employer to a charge of an "egregious" violation and the potential for criminal sanctions under the OSH Act or state criminal statutes if an employee is injured or killed as a direct result of the willful violation. Failure to abate a cited violation has the greatest cumulative monetary liability of all.

In addition to the potential civil or monetary penalties that could be assessed, OSHA regulations may be used as evidence in negligence, product liability, workers' compensation, and other actions involving employee safety and health issues.[36]

OSHA standards and regulations are the baseline requirements for safety and health that must be met, not only to achieve compliance with the OSHA regulations, but also to safeguard an organization against other potential civil actions.

DE MINIMIS VIOLATIONS

When a violation of an OSHA standard does not immediately or directly relate to safety or health, OSHA either does not issue a citation or issues a de minimis citation. Section 9 of the OSH Act provides that "[the] Secretary may prescribe procedures for the issuance of a notice in lieu of a citation with respect to de minimis violations which have no direct or immediate relationship to safety or health."[44] A de minimis notice does not constitute a citation and no fine is imposed. Additionally, there usually is no abatement period and thus there can be no violation for failure to abate.

The OSHA Compliance Field Operations Manual (OSHA Manual)[45] provides two examples of when de minimis notices are generally appropriate: (1) "In situations involving standards containing physical specificity wherein a slight deviation would not have an immediate or direct relationship to safety or health,"[46] and (2) "where the height of letters on an exit sign is not in strict conformity with the size requirements of the standard."[47]

The Occupational Safety and Health Administration has found de minimis violations in cases where employees, as well as the safety records, are persuasive in exemplifying that no injuries or lost time have been incurred.[48] Additionally, in order for OSHA to conserve valuable resources to produce a greater impact on safety and health in the workplace, it is highly likely that the secretary will encourage use of the de minimis notice in marginal cases and even in other situations where the possibility of injury is remote and potential injuries would be minor.

OTHER OR NONSERIOUS VIOLATIONS

"Other" or nonserious violations are issued where a violation could lead to an accident or occupational illness, but the probability that it would cause death or serious physical harm is minimal. Such a violation, however, does possess a direct or immediate relationship to the safety and health of workers.[49]

In distinguishing between a serious and a nonserious violation, the OSHRC has stated that "a nonserious violation is one in which there is a direct and immediate relationship between the violative condition and occupational safety and health but no such relationship that a result of injury or illness is death or serious physical harm."[51] The OSHA Manual provides guidance and examples for issuing nonserious violations. It states that

> ... an example of nonserious violation is the lack of guardrail at a height from which a fall would more probably result in only a mild sprain or cut or abrasion; i.e., something less than serious harm.[52]
>
> A citation for serious violation may be issued or a group of individual violations [which] taken by themselves would be nonserious, but together would be serious in the sense that in combination they present a substantial probability of injury resulting in death or serious physical harm to employees.[53]

A number of nonserious violations [which] are present in the same piece of equipment which, considered in relation to each other, affect the overall gravity of possible injury resulting from an accident involving the combined violations ... may be grouped in a manner similar to that indicated in the preceding paragraph, although the resulting citation will be for a nonserious violation.[54]

The difference between a serious and a nonserious violation hinges on subjectively determining the probability of injury or illness that might result from the violation. Administrative decisions have usually turned on the particular facts of the situation. The OSHRC has reduced serious citations to nonserious violations when the employer was able to show that the probability of an accident, and the probability of a serious injury or death, was minimal.[55]

SERIOUS VIOLATIONS

Section 17(k) of the OSH Act defines a serious violation as one where

...there is a substantial probability that death or serious physical harm could result from a condition which exists, or from one or more practices, means, methods, operations, or processes which have been adopted or are in use, in such place of employment unless the employer did not, and could not with

The amount of the penalty is determined by considering (1) the gravity of the violation, (2) the size of the employer, (3) the good faith of the employer, and (4) the employer's history of previous violations.[58]

To prove that a violation is within the serious category, OSHA must only show a substantial probability that a foreseeable accident would result in serious physical harm or death. Thus, contrary to common belief, OSHA does not need to show that a violation would create a high probability that an accident would result. Because substantial physical harm is the distinguishing factor between a serious and a nonserious violation, OSHA has defined "serious physical harm" as "permanent, prolonged, or temporary impairment of the body in which part of the body is made functionally useless or is substantially reduced in efficiency on or off the job." Additionally, an occupational illness is defined as "illness that could shorten life or significantly reduce physical or mental efficiency by inhibiting the normal function of a part of the body."[59]

After determining that a hazardous condition exists and that employees are exposed or potentially exposed to the hazard, the OSHA Manual instructs compliance officers to use a four-step approach to determine whether the violation is serious:[60]

1. Determine the type of accident or health hazard exposure that the violated standard is designed to prevent in relation to the hazardous condition identified.
2. Determine the type of injury or illness which it is reasonably predictable could result from the type of accident or health hazard exposure identified in step 1.

3. Determine that the type of injury or illness identified in step 2 includes death or a form of serious physical harm.
4. Determine that the employer knew or with the exercise of reasonable diligence could have known of the presence of the hazardous condition.

The OSHA Manual provides examples of serious injuries, including amputations, fractures, deep cuts involving extensive suturing, disabling burns, and concussions. Examples of serious illnesses include cancer, silicosis, asbestosis, poisoning, and hearing and visual impairment.[61]

Safety professionals should be aware that OSHA is not required to show that the employer actually knew that the cited condition violated safety or health standards. The employer can be charged with constructive knowledge of the OSHA standards. OSHA also does not have to show that the employer could reasonably foresee that an accident would happen, although it does have the burden of proving that the possibility of an accident was not totally unforeseeable. OSHA does need to prove, however, that the employer knew or should have known of the hazardous condition and that it knew there was a substantial likelihood that serious harm or death would result from an accident.[62] If the secretary cannot prove that the cited violation meets the criteria for a serious violation, the violation may be cited in one of the lesser categories.

WILLFUL VIOLATIONS

The most severe monetary penalties under the OSHA penalty structure are for willful violations. A "willful" violation can result in penalties of up to $126,749 per violation. Although the term "willful" is not defined in OSHA regulations, courts generally have defined a willful violation as "an act voluntarily with either an intentional disregard of, or plain indifference to, the Act's requirements."[63] Further, the OSHRC defines a willful violation as "action taken knowledgeably by one subject to the statutory provisions of the OSH Act in disregard of the action's legality. No showing of malicious intent is necessary. A conscious, intentional, deliberate, voluntary decision is properly described as willful."[64]

There is little distinction between civil and criminal willful violations other than the due process requirements for a criminal violation and the fact that a violation of the general duty clause cannot be used as the basis for a criminal willful violation. The distinction is usually based on the factual circumstances and the fact that a criminal willful violation results from a willful violation which caused an employee death.

According to the OSHA Manual, the compliance officer "can assume that an employer has knowledge of any OSHA violative condition of which its supervisor has knowledge; he can also presume that, if the compliance officer was able to discover a violative condition, the employer could have discovered the same condition through the exercise of reasonable diligence."[65]

Courts and the OSHRC have agreed on three basic elements of proof that OSHA must show for a willful violation. OSHA must show that the employer (1) knew or should have known that a violation existed, (2) voluntarily chose not to comply with the OSH Act to remove the violative condition, and (3) made the choice not to comply, with intentional disregard of the OSH Act's requirements or plain indifference to them properly characterized as reckless. Courts and the OSHRC have affirmed findings of willful violations in many circumstances, ranging from deliberate disregard of known safety requirements[66] through fall protection equipment not being provided.[67] Other examples of willful violations include cases where safety equipment was ordered but employees were permitted to continue work until the equipment arrived,[68] inexperienced and untrained employees were permitted to perform a hazardous job,[69] and an employer failed to correct a situation that had been previously cited as a violation.

REPEAT AND FAILURE TO ABATE VIOLATIONS

"Repeat" and "failure to abate" violations are often quite similar and confusing to human resources (HR) professionals. When, upon reinspection by OSHA, a violation of a previously cited standard is found but the violation does not involve the same machinery, equipment, process, or location, this would constitute a repeat violation. If, upon reinspection by OSHA, a violation of a previously cited standard is found but evidence indicates that the violation continued uncorrected since the original inspection, this would constitute a failure to abate violation.[70]

The most costly civil penalty under the OSH Act is for repeat violations. The OSH Act authorizes a penalty of up to $126,749 per violation. Repeat violations can also be grouped within the willful category (i.e., a willful repeat violation) to acquire maximum civil penalties.

As noted previously, a failure to abate violation occurs when, upon reinspection, the compliance officer finds that the employer has failed to take necessary corrective action and thus the violation continues uncorrected. The penalty for a failure to abate violation can be up to $12,675 per day. Safety and HR professionals should also be aware that citations for repeat violations, failure to abate violations, or willful repeat violations can be issued for violations of the general duty clause. The OSHA Manual instructs compliance officers that citations under the general duty clause are restricted to serious violations or to willful or repeat violations that are of a serious nature.[74]

CRIMINAL LIABILITY AND PENALTIES

The OSH Act provides for criminal penalties in four circumstances: (1) Giving advance notice of an inspection, without authority from the secretary; (2) intentionally falsifying a statement or OSHA record that must be prepared, maintained, or submitted under the OSH Act; (3) violation of an OSHA standard, rule, order, or regulation that causes the death of an employee; and (4) forcibly resisting or assaulting a compliance officer or other Department of Labor personnel.

The Occupational Safety and Health Administration does not have authority to impose criminal penalties directly; instead, it refers cases for possible criminal prosecution to the U.S. Department of Justice. Criminal penalties must be based on violation of a specific OSHA standard; they may not be based on a violation of the general duty clause. Criminal prosecutions are conducted like any other criminal trial, with the same rules of evidence, burden of proof, and rights of the accused. A corporation may be criminally liable for the acts of its agents or employees.[77] The statute of limitations for possible criminal violations of the OSH Act, as for other federal noncapital crimes, is five years.[78] Under federal criminal law, criminal charges may range from murder to manslaughter to conspiracy. Several charges may be brought against an employer for various separate violations under one federal indictment.

Criminal liability for a willful OSHA violation can attach to an individual or a corporation. In addition, corporations may be held criminally liable for the actions of their agents or officials.[79] Safety professionals and other corporate officials may also be subject to criminal liability under a theory of aiding and abetting the criminal violation in their official capacity with the corporation.[80]

Safety professionals should also be aware that an employer could face two prosecutions for the same OSHA violation without the protection of double jeopardy. The OSHRC can bring an action for a civil willful violation using the monetary penalty structure described previously and the case then may be referred to the Justice Department for criminal prosecution of the same violation.[81]

Prosecution of willful criminal violations by the Justice Department has been rare in comparison to the number of inspections performed and violations cited by OSHA on a yearly basis. However, the use of criminal sanctions has increased substantially in the last few years. With adverse publicity being generated as a result of workplace accidents and deaths[82] and Congress emphasizing reform, a decrease in criminal prosecutions is unlikely.

The law regarding criminal prosecution of willful OSH Act violations is still emerging. Although few cases have actually gone to trial, in most situations the mere threat of criminal prosecution has encouraged employers to settle cases with assurances that criminal prosecution would be dismissed. Many state plan states are using criminal sanctions permitted under their state OSH regulations more frequently.[83] State prosecutors have also allowed use of state criminal codes for workplace deaths.[84]

After a disaster situation, especially if a fatality is involved, safety professionals should exercise extreme caution. The potential for criminal sanctions and criminal prosecution is substantial if a willful violation of a specific OSHA standard is directly involved in the death. The OSHA investigation may be conducted from a criminal perspective in order to gather and secure the appropriate evidence to later pursue criminal sanctions.[85] A prudent safety professional facing a workplace fatality investigation should address the OSHA investigation with legal counsel present and reserve all rights guaranteed under the U.S. Constitution.[86] Obviously, under no circumstances should a safety professional condone or attempt to conceal facts or evidence, which consists of a cover-up.

INSPECTIONS AND CITATIONS

The Occupational Safety and Health Administration performs all enforcement functions under the OSH Act. Under Section 8(a) of the Act, OSHA compliance officers have the right to enter any workplace of a covered employer without delay, to inspect and investigate a workplace during regular hours and at other reasonable times, and to obtain an inspection warrant if access to a facility or operation is denied.[87] Upon arrival at an inspection site, the compliance officer is required to present his or her credentials to the owner or designated representative of the employer before starting the inspection. The employer representative and an employee or union representative may accompany the compliance officer on the inspection. Compliance officers can question the employer and employees and inspect required records, such as the OSHA Form 200, which records injuries and illnesses.[88] Most compliance officers cannot issue on-the-spot citations; they only have authority to document potential hazards and report or confer with the OSHA area director before issuing a citation.

A compliance officer or any other employee of OSHA may not provide advance notice of the inspection under penalty of law.[89] The OSHA area director is, however, permitted to provide notice under the following circumstances:[90]

1. In cases of apparent imminent danger, to enable the employer to correct the danger as quickly as possible.
2. When the inspection can most effectively be conducted after regular business hours or where special preparations are necessary.
3. To ensure the presence of employee and employer representatives or appropriate personnel needed to aid in inspections.
4. When the area director determines that advance notice would enhance the probability of an effective and thorough inspection.

Compliance officers can also take environmental samples and obtain photographs related to the inspection. Additionally, compliance officers, can use other "reasonable investigative techniques," including personal sampling equipment, dosimeters, air-sampling badges, and other equipment.[91] Compliance officers must, however, take reasonable precautions when using photographic or sampling equipment to avoid creating hazardous conditions (i.e., a spark-producing camera flash in a flammable area) or disclosing a trade secret.[92]

An OSHA inspection has four basic components: (1) The opening conference, (2) the walk-through inspection, (3) the closing conference, and (4) the issuance of citations, if necessary. In the opening conference, the compliance officer may explain the purpose and type of inspection to be conducted, request records to be evaluated, question the employer, ask for appropriate representatives to accompany him or her during the walk-through inspection, and ask additional questions or request more information. The compliance officer may, but is not required to, provide the employer with copies of the applicable laws and regulations governing procedures and health and safety standards. The opening conference is usually brief and informal, its primary purpose being to establish the scope and purpose of the walk-through inspection.

After the opening conference and review of appropriate records, the compliance officer, usually accompanied by a representative of the employer and a representative of the employees, conducts a physical inspection of the facility or worksite.[93] The general purpose of this walk-through inspection is to determine whether the facility or worksite complies with OSHA standards. The compliance officer must identify potential safety and health hazards in the workplace, if any, and document them to support issuance of citations.[94]

When the walk-through inspection is completed, the compliance officer usually conducts an informal meeting with the employer or the employer's representative to "informally advise (the employer) of any apparent safety or health violations disclosed by the inspection.[96] The compliance officer informs the employer of the potential hazards observed and indicates the applicable section of the standards allegedly violated, advises that citations may be issued, and informs the employer or representative of the appeal process and rights.[97] Additionally, the compliance officer advises the employer that the OSH Act prohibits discrimination against employees or others for exercising their rights.[98]

In an unusual situation, the compliance officer may issue a citation(s) on the spot. When this occurs, the compliance officer informs the employer of the abatement period, in addition to the other information provided at the closing conference. In most circumstances, the compliance officer will leave the workplace and file a report with the area director who has authority, through the Secretary of Labor, to decide whether a citation should be issued, compute any penalties to be assessed, and set the abatement date for each alleged violation. The area director, under authority from the secretary, must issue the citation with "reasonable promptness."[99] Citations must be issued in writing and must describe with particularity the violation alleged, including the relevant standard and regulation. There is a six-month statute of limitations, and the citation must be issued or vacated within this time period. OSHA must serve notice of any citation and proposed penalty by certified mail, unless there is personal service, to an agent or officer of the employer.[100]

After the citation and notice of proposed penalty are issued, but before the notice of contest by the employer is filed, the employer may request an informal conference with the OSHA area director. The general purpose of the informal conference is to clarify the basis for the citation, modify abatement dates or proposed penalties, seek withdrawal of a cited item, or otherwise attempt to settle the case. This conference, as its name implies, is an informal meeting between the employer and OSHA. Employee representatives must have an opportunity to participate if they so request. Safety professionals should note that the request for an informal conference does not "stay" (delay) the 15-working-day period to file a notice of contest to challenge the citation.[101]

Under the OSH Act, an employer, employee, or authorized employee representative (including a labor organization) is given 15 working days from when the citation is issued to file a "notice of contest." If a notice of contest is not filed within 15 working days, the citation and proposed penalty become a final order of the Occupational Safety and Health Review Commission (OSHRC) and are not subject to review by any court or agency. If a timely notice of contest is filed in good faith, the abatement requirement is tolled (temporarily suspended or delayed) and a hearing is scheduled. The employer also has the right to file a petition for modification of the abatement period (PMA) if the employer is unable to comply with the abatement period provided in the citation. If OSHA contests the PMA, a hearing is scheduled to determine whether the abatement requirements should be modified.

When the notice of contest by the employer is filed, the secretary must immediately forward the notice to the OSHRC, which then schedules a hearing before its administrative law judge (ALJ). The Secretary of Labor is labeled the "complainant," and the employer the "respondent." The ALJ may affirm, modify, or vacate the citation, any penalties, or the abatement date. Either party can appeal the administrative law judge's decision by filing a petition for discretionary review (PDR). Additionally, any member of the OSHRC may "direct review" of any decision by an ALJ, in whole or in part, without a PDR. If a PDR is not filed and no member of the OSHRC directs a review, the decision of the ALJ becomes final in 30 days. Any party may appeal a final order of the OSHRC by filing a petition for review in the U.S. Court of Appeals for the circuit in which the violation is alleged to have occurred or in the U.S. Court of Appeals for the District of Columbia Circuit. This petition for review must be filed within 60 days from the date of the OSHRC's final order.

OSHA INSPECTION CHECKLIST

The following is a recommended checklist for safety professionals in order to prepare for an OSHA inspection in general or after a disaster situation:[102]

1. Assemble a team from the management group and identify specific responsibilities in writing for each team member. The team members should be given appropriate training and education and should include, but not be limited to
   a. OSHA inspection team coordinator
   b. Document-control individual
   c. Individuals to accompany the OSHA inspector
   d. Media coordinator
   e. Accident investigation team leader (where applicable)
   f. Notification person
   g. Legal advisor (where applicable)
   h. Law enforcement coordinator (where applicable)
   i. Photographer
   j. Industrial hygienist
2. Decide on and develop a company policy and procedures to provide guidance to the OSHA inspection team.
3. Prepare an OSHA inspection kit, including all equipment necessary to properly document all phases of the inspection. The kit should include equipment such as a camera (with extra film and batteries), a tape player (with extra batteries), a video camera, pads, pens, and other appropriate testing and sampling equipment (e.g., noise-level meter, air-sampling kit, etc.).
4. Prepare basic forms to be used by the inspection team members during and following the inspection.
5. When notified that an OSHA inspector has arrived, assemble the team members along with the inspection kit.

6. Identify the inspector. Check his or her credentials, and determine the reason for and type of inspection to be conducted.
7. Confirm the reason for the inspection with the inspector (targeted, routine inspection, accident, or in response to a complaint).
   a. For a random or target inspection:
      i. Did the inspector check the OSHA Injury and Illness Records (300 log, 300A, 301s)?
      ii. Was a warrant required?
   b. For an employee complaint inspection:
      i. Did inspector have a copy of the complaint? If so, obtain a copy.
      ii. Do allegations in the complaint describe an OSHA violation?
      iii. Was a warrant required?
      iv. Was the inspection protested in writing?
   c. For an accident investigation inspection:
      i. How was OSHA notified of the accident?
      ii. Was a warrant required?
      iii. Was the inspection limited to the accident location?
   d. If a warrant is presented:
      i. Were the terms of the warrant reviewed by local counsel?
      ii. Did the inspector follow the terms of the warrant?
      iii. Was a copy of the warrant acquired?
      iv. Was the inspection protested in writing?
8. The opening conference:
   a. Who was present?
   b. What was said?
   c. Was the conference taped or otherwise documented?
9. Records:
   a. What records were requested by the inspector?
   b. Did the document-control coordinator number the photocopies of the documents provided to the inspector?
   c. Did the document control coordinator maintain a list of all photocopies provided to the inspector?
10. Facility inspection:
    a. What areas of the facility were inspected?
    b. What equipment was inspected?
    c. Which employees were interviewed?
    d. Who was the employee or union representative present during the inspection?
    e. Were all the remarks made by the inspector documented?
    f. Did the inspector take photographs?
    g. Did a team member take similar photographs?[102]

In a situation where local or state prosecutors (i.e., district attorney, or DA) respond to a fatality or disaster situation, safety professionals should exercise extreme caution and acquire legal representation prior to providing any information or documents. The prosecutor is usually investigating from a criminal perspective and would utilize the state criminal code as the basis for any charges. Safety professionals should preserve all constitutional rights, including the right to remain silent and the right to counsel in these situations. As noted in the Miranda warnings provided by law enforcement after an arrest, "You have the right to remain silent; everything you say can and will be used against you in a court of law; you have the right to counsel ..."

In summation, being prepared and knowing the rules can assist safety professionals in protecting their organization's assets during any type of governmental regulatory inspection. Although we have used OSHA as an example due to the fact that most safety professionals interact with OSHA more than other governmental agencies, safety professionals with responsibilities in human resources, environmental, and other areas should have an in-depth familiarity with the regulatory scheme and rules of the specific governmental entity. Preparation and knowledge will permit smooth sailing during any inspection!

ENDNOTES

1. 29 C.F.R. §1975.3(6).
2. *Id.* at §652(7).
3. See, e.g., Atomic Energy Act of 1954, 42 U.S.C. §2021.
4. 29 C.F.R. §1975(6).
5. 29 U.S.C.A. §652(5) (no coverage under OSH Act, when U.S. government acts as employer).
6. *Id.*
7. See, e.g., *Navajo Forest Prods. Indus.*, 8 O.S.H. Cases 2694 (OSH Rev. Comm'n 1980), aff'd, 692 F.2d 709, 10 O.S.H. Cases 2159.
8. 29 U.S.C.A. §654(a)(1).
9. In Section 18(b), the OSH Act provides that any state "which, at any time, desires to assume responsibility for development and the enforcement therein of occupational safety and health standards relating to any ... issue with respect to which a federal standard has been promulgated ... shall submit a state plan for the development of such standards and their enforcement."
10. *Id.* at §667(c). After an initial evaluation period of a least three years during which OSHA retains concurrent authority, a state with an approved plan gains exclusive authority over standard settings, inspection procedures, and enforcement of health and safety issues covered under the state plan. See also *Noonan v. Texaco*, 713 P.2d 160 (Wyo. 1986); Plans for the Development and Enforcement of State Standards, 29 C.F.R. §667(f) (1982) and §1902.42(c)(1986). Although the state plan is implemented by the individual state, OSHA continues to monitor the program and may revoke the state authority if the state does not fulfill the conditions and assurances contained within the proposed plan.
11. Some states incorporate federal OSHA standards into their plans and add only a few of their own standards as a supplement. Other states, such as Michigan and California, have added a substantial number of separate and independently promulgated standards. See, generally, Employee Safety and Health Guide (CCH) §§5000–5840 (1987) (compiling all state plans). Some states also add their own penalty structures. For example, under Arizona's plan, employers may be fined up to $150,000 and sentenced to one-and-one-half years in prison for knowing violations of state standards that cause death to an employee and may also have to pay $25,000 in compensation to the victim's family. If the employer is a corporation, the maximum fine is $1 million. See Ariz. Rev. Stat. Ann. §§13-701, 13-801, 23-4128, 23-418.01, 13-803 (Supp. 1986).

12. For example, under Kentucky's state plan regulations for controlling hazardous energy (i.e., lockout/tagout), locks would be required rather than locks or tags being optional, as under the federal standard. Lockout/tagout is discussed in more detail in Chapter 2.

13. 29 U.S.C. §667.

14. 29 U.S.C.A. §667; 29 C.F.R. §1902.

15. The states and territories operating their own OSHA programs are Alaska, Arizona, California, Hawaii, Indiana, Iowa, Kentucky, Maryland, Michigan, Minnesota, Nevada, New Mexico, North Carolina; partial federal OSHA enforcement, Oregon, Puerto Rico, South Carolina, Tennessee, Utah, Vermont, Virgin Islands, Virginia, Washington, and Wyoming.

16. 29 U.S.C. §655(b).

17. 29 U.S.C.A. §655(b)(5).

18. 29 U.S.C. 1910.

19. 29 C.F.R. §1911.15. By regulation, the Secretary of Labor has prescribed more detailed procedures than the OSH Act specifies to ensure participation in the process of setting new standards.

20. 29 U.S.C. §1910.

21. 29 U.S.C. §655(b).

22. 5 U.S.C. §553.

23. 29 C.F.R. §1911.15.

24. *Taylor Diving & Salvage Co. v. Department of Labor*, 599 F.2d 622 7 O.S.H. Cases 1507 (5th Cir. 1979).

25. 29 U.S.C. §655(c).

26. *Id.* at §655(b)(1).

27. *Id.* at §656(a)(1). NACOSH was created by the OSH Act to "advise, consult with, and make recommendations ... on matters relating to the administration of the Act." Normally, for new standards, the secretary has established continuing committees and ad hoc committees to provide advice regarding particular problems or proposed standards.

28. *OSHA Compliance Field Operations Manual (OSHA Manual)* at XI-C3c (Apr. 1977).

29. *Id.*

30. *Id.* at XI-C3c(2).

31. *Id.* at (3)(a).

32. 29 U.S.C. §666.

33. *Id.* at §666(b).

34. For example, if a company possesses 25 identical machines, and each of these machines is found to have the identical serious violation, this would theoretically constitute 25 violations, rather than one violation on 25 machines, and a possible monetary fine of $175,000, rather than a maximum of $7,000.

35. *Occupational Safety & Health Reporter*, V. 23, No. 32, Jan. 12, 1994.

36. See *infra* at §1.140.

37. 29 U.S.C. §§658(a), 666(c).

38. *Id.* at §666(j).

39. *Id.* at §666(c).

40. *Id.* at §666(a).

41. *Id.*

42. *Id.* at (b).

43. *Id.* at (e).

44. *Id.* at §658(a).

45. *Supra* at note 62.

46. *Id.* at VII-B3a.

47. *Id.*

48. *Hood Sailmakers*, 6 O.S.H. Cases 1207 (1977).

49. *OSHA Manual, supra* at note 62, at VIII-B2a. The proper nomenclature for this type of violation is "other" or "other than serious." Many safety and health professionals classify this type of violation as nonserious for explanation and clarification purposes.

50. A nonserious penalty is usually less than $100 per violation.

51. *Crescent Wharf & Warehouse Co.*, 1 O.S.H. Cases 1219, 1222 (1973).

52. *OSHA Manual, supra* at note 62, at VIII-B2a.

53. *Id.* at B2b(1).

54. *Id.* at (2).

55. See *Secretary v. Diamond Indus.*, 4 O.S.H. Cases 1821 (1976); *Secretary v. Northwest Paving*, 2 O.S.H. Cases 3241 (1974); *Secretary v. Sky-Hy Erectors & Equip.*, 4 O.S.H. Cases 1442 (1976). But see *Shaw Constr. v. OSHRC*, 534 F.2d 1183, 4 O.S.H. Cases 1427 (5th Cir. 1976) (holding that serious citation was proper whenever accident was merely possible.)

56. 29 U.S.C. §666(j).

57. *Id.*

58. 29 U.S.C. §666(i).

59. *OSHA Manual, supra* at note 62, at IV-B-1(b)(3)(a),(c).

60. *Id.* at VIII-B1b(2)(c). In determining whether a violation constitutes a serious violation, the compliance officer is functionally describing the *prima facie* case that the secretary would be required to prove, i.e., (1) the causal link between the violation of the safety or health standard and the hazard, (2) reasonably predictable injury or illness that could result, (3) potential of serious physical harm or death, and (4) the employer's ability to foresee such harm by using reasonable diligence.

61. *Id.* at VIII-B1c(3)a.

62. *Id.* at (4). See also *Cam Indus.*, 1 O.S.H. Cases 1564 (1974); *Secretary v. Sun Outdoor Advertising*, 5 O.S.H. Cases 1159 (1977).

63. *Cedar Constr. Co. v. OSHRC*, 587 F.2d 1303, 6 O.S.H. Cases 2010, 2011 (D.C. Cir. 1971). Moral turpitude or malicious intent are not necessary elements for a willful violation. *U.S. v. Dye Constr.*, 522 F.2d 777, 3 O.S.H. Cases 1337 (4th Cir. 1975); *Empire-Detroit Steel v. OSHRC*, 579 F.2d 378, 6 O.S.H. Cases 1693 (6th Cir. 1978).

64. *P.A.F. Equip. Co.*, 7 O.S.H. Cases 1209 (1979).

65. *OSHA Manual, supra* at note 62, at VIII-B1c(4).

66. *Universal Auto Radiator Mfg. Co. v. Marshall*, 631 F.2d 20, 8 O.S.H. Cases 2026 (3d Cir. 1980).

67. *Haven Steel Co. v. OSHRC*, 738 F.2d 397, 11 O.S.H. Cases 2057 (10th Cir. 1984).

68. *Donovan v. Capital City Excavating Co.*, 712 F.2d 1008, 11 O.S.H. Cases 1581 (6th Cir. 1983).

69. *Ensign-Bickford Co. v. OSHRC*, 717 F.2d 1419, 11 O.S.H. Cases 1657 (D.C. Cir. 1983).

70. *OSHA Manual, supra* at note 62, at VIII-B5c.

71. *Id.* at IV-B5(c)(1).

72. *Id.* at VIII-B5d.

73. *Id.*

74. *Id.* at XI-C5c.

75. 29 U.S.C. §666(e)–(g). See also *OSHA Manual, supra* note 62 at VI-B.

76. A repeat criminal conviction for a willful violation causing an employee death doubles the possible criminal penalties.

77. 29 C.F.R. §5.01(6).

78. *U.S. v. Dye Const. Co.*, 510 F.2d 78, 2 O.S.H. Cases 1510 (10th Cir. 1975).

79. *U.S. v. Crosby & Overton*, No. CR-74-1832-F (S.D. Cal. Feb. 24, 1975).

80. 18 U.S.C. §2.

81. These are uncharted waters. Employers may argue due process and double jeopardy, but OSHA may argue that it has authority to impose penalties in both contexts. There are currently no cases on this issue.

82. Jefferson, Dying for work, *A.B.A.*, #J. 46 (Jan.), 1993.

83. See, e.g., Levin, Crimes against employees: substantive criminal sanctions under the Occupational Safety and Health Act, *Am. Crim. L. Rev.*, 14, 98, 1977.

84. See Chapter 5.

85. See L.A. law: prosecuting workplace killers, *A.B.A.*, #J. 48. (Los Angeles prosecutor's "roll out" program could serve as model for OSHA.)

86. See 29 C.F.R. §1903.8.

87. See *infra* §§1.10 and 1.12.

88. 29 C.F.R. §1903.8.

89. 29 U.S.C. §17(f). The penalty for providing advance notice, upon conviction, is a fine of not more than $1000, imprisonment for not more than 6 months, or both.

90. *Occupational Safety and Health Law*, 208–209 (1988).

91. 29 C.F.R. §1903.7(b) (revised by 47 *Fed. Reg.* 5548 [1982].)

92. See, e.g., 29 C.F.R. §1903.9. Under §15 of the OSH Act, all information gathered or revealed during an inspection or proceeding that may reveal a trade secret as specified under 18 U.S.C. §1905 must be considered confidential, and breach of that confidentiality is punishable by a fine of not more than $1000, imprisonment of not more than one year, or both, and removal from office or employment with OSHA.

93. It is highly recommended by the authors that a company representative accompany the OSHA inspection during the walk-through inspection.

94. *OSHA Manual, supra* note 62, at III-D8.

95. The OSHA 1 Inspection Report Form includes the following:
    - Establishment's name
    - Inspection number
    - Type of legal entity
    - Type of business or plant
    - Additional citations
    - Names and addresses of all organized employee groups
    - Authorized representative of employees
    - Employee representative contacted
    - Other persons contacted
    - Coverage information (state of incorporation, type of goods or services in interstate commerce, etc.)
    - Date and time of entry
    - Date and time that the walk-through inspection began
    - Date and time closing conference began
    - Date and time of exit
    - Whether a follow-up inspection is recommended
    - Compliance officer's signature and date
    - Names of other compliance officers,
    - Evaluation of safety and health programs (checklist)
    - Closing conference checklist
    - Additional comments

96. 29 C.F.R. §1903.7(e).

97. *OSHA Manual, supra* at note 62, at III-D9.

98. 29 U.S.C. §660(c)(1).

99. *Id.* at §658.

100. Fed. R. Civ. P. 4(d)(3).

101. 29 U.S.C. §659(a).

102. Schneid, T., Preparing for an OSHA inspection, *Kentucky Manufacturer*, February 1992.

Appendix H: Machine Guarding Checklist

Answering the following questions should help the interested reader determine the safeguarding needs of his/her own workplace by drawing attention to hazardous conditions or practices requiring correction.

REQUIREMENTS FOR ALL SAFEGUARDS

		Yes	No
1.	Do the safeguards provided meet the minimum OSHA requirements?	___	___
2.	Do the safeguards prevent workers' hands, arms, and other body parts from making contact with dangerous moving parts?	___	___
3.	Are the safeguards firmly secured and not easily removable?	___	___
4.	Do the safeguards ensure that no object will fall into the moving parts?	___	___
5.	Do the safeguards permit safe, comfortable, and relatively easy operation of the machine?	___	___
6.	Can the machine be oiled without removing the safeguard?	___	___
7.	Is there a system for shutting down the machinery before safeguards are removed?	___	___
8.	Can the existing safeguards be improved?	___	___

MECHANICAL HAZARDS

POINT OF OPERATION

		Yes	No
1.	Is there a point-of-operation safeguard provided for the machine?	___	___
2.	Does it keep the operator's hands, fingers, and body out of the danger area?	___	___
3.	Is there evidence that the safeguards have been tampered with or removed?	___	___
4.	Could you suggest a more practical, effective safeguard?	___	___
5.	Could changes be made on the machine to eliminate the point-of-operation hazard entirely?	___	___

POWER TRANSMISSION APPARATUS

		Yes	No
1.	Are there any unguarded gears, sprockets, pulleys, or flywheels on the apparatus?	___	___
2.	Are there any exposed belts or chain drivers?	___	___
3.	Are there any exposed set screws, key ways, collars, etc.?	___	___
4.	Are starting and stopping controls within easy reach of the operator?	___	___
5.	If there is more than one operator, are separate controls provided?	___	___

OTHER MOVING PARTS

		Yes	No
1.	Are safeguards provided for all hazardous moving parts of the machine, including auxiliary parts?	___	___

NONMECHANICAL HAZARDS

		Yes	No
1.	Have appropriate measures been taken to safeguard workers against noise hazards?	___	___
2.	Have special guards, enclosures, or personal protective equipment been provided, where necessary, to protect workers from exposure to harmful substances used in machine operation?	___	___

ELECTRICAL HAZARDS

		Yes	No
1.	Is the machine installed in accordance with National Fire Protection Association and National Electrical Code requirements?	___	___
2.	Are there loose conduit fittings?	___	___
3.	Is the machine properly grounded?	___	___
4.	Is the power supply correctly fused and protected?	___	___
5.	Do workers occasionally receive minor shocks while operating any of the machines?	___	___

TRAINING

	Yes	No
1. Do operators and maintenance workers have the necessary training in how to use the safeguards and why?	____	____
2. Have operators and maintenance workers been trained in where the safeguards are located, how they provide protection, and what hazards they protect against?	____	____
3. Have operators and maintenance workers been trained in how and under what circumstances guards can be removed?	____	____
4. Have workers been trained in the procedures to follow if they notice guards that are damaged, missing, or inadequate?	____	____

PROTECTIVE EQUIPMENT AND PROPER CLOTHING

	Yes	No
1. Is protective equipment required?	____	____
2. If protective equipment is required, is it appropriate for the job, in good condition, kept clean and sanitary, and stored carefully when not in use?	____	____
3. Is the operator dressed safely for the job (e.g., no loose-fitting clothing or jewelry)?	____	____

MACHINERY MAINTENANCE AND REPAIR

	Yes	No
1. Have maintenance workers received up-to-date instruction on the machines they service?	____	____
2. Do maintenance workers lock out the machine from its power sources before beginning repairs?	____	____
3. Where several maintenance persons work on the same machine, are multiple lockout devices used?	____	____
4. Do maintenance personnel use appropriate and safe equipment during their repair work?	____	____
5. Is the maintenance equipment itself properly guarded?	____	____
6. Are maintenance and servicing workers trained in the requirements of 29 C.F.R. 1910.147 (lockout/tagout of hazards), and do the procedures for lockout/tagout exist before they attempt their tasks?	____	____

Index

Page numbers followed by f and t indicate figures and tables, respectively.

Milton Keynes UK
Ingram Content Group UK Ltd.
UKHW031144141024
449569UK00024B/1071